SpringerBriefs in Philosophy

More information about this series at http://www.springer.com/series/10082

Nicolas Clerbout · Shahid Rahman

Linking Game-Theoretical Approaches with Constructive Type Theory

Dialogical Strategies, CTT Demonstrations and the Axiom of Choice

Springer

Nicolas Clerbout
Instituto de Filosofía; CDHACS
Universidad de Valparaíso
Valparaíso
Chile

Shahid Rahman
UMR-CNRS 8163: STL
University of Lille III
Villeneuve d'Ascq
France

ISSN 2211-4548 ISSN 2211-4556 (electronic)
SpringerBriefs in Philosophy
ISBN 978-3-319-19062-4 ISBN 978-3-319-19063-1 (eBook)
DOI 10.1007/978-3-319-19063-1

Library of Congress Control Number: 2015939429

Springer Cham Heidelberg New York Dordrecht London

Printed on acid-free paper

Springer International Publishing AG Switzerland is part of Springer Science+Business Media
(www.springer.com)

To Göran Sundholm and Gerhard Heinzmann

Preface

A brief examination of the most recent literature in logic shows that a host of research in this area studies the interface between games, logic, and epistemology. These studies provide the basis for ongoing enquiries in the history and philosophy of logic, going from the Indian, the Greek, the Arabic traditions, the Obligationes of the Middle Ages to the most contemporary developments in the fields of theoretical computer science, computational linguistics, artificial intelligence, social sciences, and legal reasoning. In fact, a dynamic turn, as Johan van Benthem puts it, is taking place where the epistemic aspects of inference are linked with game theoretical approaches to meaning.[1]

This turn came about in the 1960s when Paul Lorenzen and Kuno Lorenz developed dialogical logic—inspired by Wittgenstein's language games and mathematical game theory—and when some time later Jaakko Hintikka combined game-theoretical semantics with epistemic (modal) logic. If we had to pinpoint a date, it would be 1958 with Lorenzen's talk[2] "Logik und Agon."

However, these two approaches to logic—the dialogical one and the one based on Hintikka's GTS—springing from a dynamic reading of the epistemic conception of logic, disregarded a major advance precisely in the epistemic approach to logic, namely the development by Per Martin-Löf of Constructive Type Theory (CTT)—with the sole exception of the pioneering paper of Ranta (1988).[3] This framework,

[1]New results in linear logic by J.-Y. Girard at the interface between mathematical game theory and proof theory on the one hand and argumentation theory and logic on the other resulted in the work of, among others, S. Abramsky, J. van Benthem, A. Blass, H. van Ditmarsch, D. Gabbay, M. Hyland, W. Hodges, G. Japaridze, E. Krabbe, H. Prakken, G. Sandu, D. Walton and J. Woods, who placed game semantics at the center of a new concept of logic in which logic is understood as a dynamic instrument of inference. In this context see also Blass (1992), Abramsky and Mellies (1999), Girard (1999), Lecomte and Quatrini (2010, 2011), Lecomte (2011), Lecomte and Trançon (2011).

[2]Published as Lorenzen (1960).

[3]We will discuss further on the differences with our own approach, however, let us point out that Ranta's paper was written from the CTT perspective on the game theoretical approach to meaning, rather than the other way round.

providing a type theoretical development of the Curry-Howard-isomorphism and introducing dependent-types, leads to the formulation of a fully-interpreted language—a language with content challenging the standard metalogical approach to meaning of model theoretic semantics in general and of the modal-interpretation of epistemic logic in particular. Furthermore, an inferential and contentual language based on CTT (Sundholm 1986, 2001; Ranta 1994) has now been successfully applied not only to the semantics of natural languages but also to the foundations of logic, computer sciences, and constructive mathematics. Philosophically speaking, CTT shares the Kantian view that judgements, rather than propositions, constitute the foundation of knowledge. According to this perspective, the basic ontology is determined by the two fundamental forms of judgement, namely categorical judgements with *independent* proof-objects and hypothetical judgements with *dependent* proof-objects (i.e., functions). See Chap. 1.

Interestingly enough CTT supplies, as discussed by Rahman and Clerbout (2013), the basis for research on the same areas that the dialogical framework—after some initial analyses—stopped to work on, namely the foundations of mathematics and the development of a general dialogical theory of meaning. Up to now, the lack of interface between the game theoretical approaches and CTT is particularly striking because of their common philosophical ground. Let us very briefly expound on this. One way to put it is to follow Mathieu Marion's[4] suggestion to use Brandom's (1994, 2000) pragmatist take on inferentialism which is led by three insights, two of Kantian origin on the one hand and one stemming from Brandom's reading of Hegel on the other hand:

- that judgements are the fundamental units of knowledge,
- that human cognition and action are characterized by certain kinds of normative assessment,[5] and

[4]In fact, Marion (2006, 2009, 2010) was the first to propose a link between Brandom's pragmatist inferentialism and dialogue logic in the context of Wilfried Hodges challenging to the game-theoretical approaches (2001, 2004). Another relevant antecedent of the present work is the Ph.D. dissertation of Keiff (2006–2009) providing a thorough formulation of the dialogical approach within the framework of speech-act theory.

[5]The normative aspect, established on the shift from *Cartesian certainty* to *bindingness of rules* distinguishes Brandom's pragmatism from others':

> One of the strategies that guided this work is a commitment to the fruitfulness of shifting theoretical attention from the Cartesian concern with the grip we have on concepts—for Descartes, in the particular form of the centrality of the notion of **certainty** (…)—to the Kantian concern with the grip concepts have on us, that is, the notion of **necessity** as the bindingness of the rules (including inferential ones) that determine how it is correct to apply those concepts. (Brandom 1994, p. 636)

- that communication is mainly conceived as cooperation in a joint social activity rather than as sharing contents.[6]

As mentioned above, the key aspect of the epistemic approach is that assertion or judgement amounts to a knowledge claim, independently of classical or intuitionistic views—cf. Prawitz (2012, p. 47). So if an expression's meaning stems from its role in assertions, then the approach to meaning is epistemic. According to Brandom, the normative aspect is implemented through W. Sellar's notion of games of giving and asking for reasons, games that deploy the way commitments and entitlements are intertwined. Indeed, according to Brandom's view, it is the chain of commitments and entitlements in a game of giving and asking for reasons that binds up judgement with inference.[7] Sundholm (2013) provides the following formulation of the notion of inference in a communicative context which can be seen as a core description of Brandom's pragmatist inferentialism:[8]

> When I say "Therefore" **I give others my authority for asserting the conclusion, given theirs for asserting the premisses.**

This is quite close to the main tenet of the dialogical approach to meaning, though with one crucial difference. The pragmatist approach to meaning of the dialogical framework not only endorses Brandom's pragmatist inferentialism claim that the meaning of linguistic expressions is related to their role in games of questions and answers, but it also endorses Brandom's notion of justification of a judgement as involving the interaction of commitments and entitlements. The important difference is that dialogicians insist that more fundamental lower levels should be distinguished—as discussed in Chap. 2. These lower semantic levels include (i) the description of how to formulate a suitable question for a given posit and how to answer it, and (ii) the development of plays, composed of several combinations of sequences of questions and answers brought forward as responses to the posit of a thesis. From the dialogical perspective, the level of judgements corresponds to the final stage of the chain of interactions just mentioned. More

---

[6] In relation to the model of holistic communication which is considered here, Brandom writes:

> Holism about inferential significances has different theoretical consequences depending on whether one thinks of communication in terms of **sharing** a relation to one and the same **thing** (grasping a common meaning) or in terms of **cooperating** in a joint **activity** (…) (Brandom 1994, p. 479).

[7] Moreover, according to Brandom, games of asking for reasons and giving them constitute the basis of any linguistic practice:

> Sentences are expressions whose unembedded utterance performs a speech act such as making a claim, asking a question, or giving a command. Without expressions of this category, there can be no speech acts of any kind, and hence no specifically linguistic practice. (Brandom 2000, p. 125)

[8] Actually, Sundholm bases his formulation on J.L. Austin's remarks in his celebrated paper "Other Minds" (Austin 1946) rather than on Brandom's work.

precisely, the justifications of judgements correspond to the level of winning strategies, selecting those plays that turn out to be relevant for the drawing of inferences. Furthermore, as our discussion of the Axiom of Choice (Chap. 4) shows, the game theoretical take on dependent types is rooted on choice dependences, which can be seen as a result of intertwining games of questions and answers.

Let us point out that the distinctions drawn within the dialogical framework between local meaning, play level and strategy level, which we will explain in detail in Chap. 2, seem to provide an answer to Brandom's problem that the "grasp of concepts" amounts to the mastery of inferential roles but this

> (…) does not mean that in order to count as grasping a particular concept an individual must be disposed to make or otherwise endorse in practice all the right inferences involving it. To be in the game at all, one must make **enough** of the right moves—but how much is enough is quite flexible. (Brandom 1994, p. 636)

Indeed, from the dialogical point of view, in order to grasp the meaning of an expression, the individual does not need to know the moves ensuring his victory (he must not have a winning strategy) and does not even need to win at all. What is required is that he knows what the relevant moves are which he is entitled and committed to (local meaning) in order to develop a play. In a similar way, knowing how to play chess does not necessarily mean to actually be in possession of a winning strategy. Knowing how to play allows to know what can count as a winning strategy, when there is one: strategic legitimacy (*Geltung*) is not to be found at the level of meaning-explanation. Thus, one way to see the motivations for proposing to link CTT and games is to give the technical elements binding the pragmatist approach that grasps concepts in Brandom's style with the proof-theoretical CTT take on meaning.

At this point of the discussion, we hope that the grounds for working out systematically the links between game theoretical approaches and CTT—or at least a glimpse of them—should be clear enough. A pending task, that we will not undertake here, is to discuss how the rigorous elaboration of a fully-interpreted language in terms of CTT fits with Brandom's pragmatic inferentialism. However, as stressed in Chap. 4, a fully-interpreted language lies at the core of the dialogical take on meaning.[9] More generally and summing up, the development of a dialogical approach to Constructive Type Theory can be motivated by the following considerations:

1. In his book *The Interactive Stance*, Ginzburg (2012) stresses the utmost importance of taking conversational (interactive) aspects into account in order to develop a theory of meaning, in which meaning is constituted during the interaction. In order to implement such a theory of meaning Ginzburg uses CTT in which the so-called "metalogica" rules constituting meaning are explicitly imported into the object language. Moreover, Ginzburg designed some kind of

---

[9]For a thorough discussion on the subject see Rahman and Clerbout (2013).

language games called dialogical-gameboards in order to capture the dynamic aspects of everyday dialogues.

Now, if we take seriously the claim that meaning is constituted by and within interaction, then we expect that the semantics of the underlying logical elements is also understood dialogically. In this context, a dialogical approach to Constructive Type Theory provides both a dialogical framework for the underlying logic and a natural link to the dialogical gameboards.

2. In some recent work, Dybjer (2012) proposed to study the relation between the intensional and the extensional notions of identity, defined in CTT by game theoretical means. Once more, in this context an approach to CTT that is already game-theoretical seems to be desirable.
3. The links between game-theoretical approaches and CTT seem to be very natural. Indeed, the CTT approach is very natural to the dialogical framework in which the meaning of the logical constants is given by moves such as challenges and choices constituting the explicit development of a play. Challenges and choices are in fact part of the object language of the development of a play.
4. The fathers of the dialogical framework had the project of developing both a general theory of meaning (and not only of logic) and to open a new path on the foundations of (constructive) mathematics and logic. Here, too, it is CTT that enriches the dialogical approach.[10]

Given these motivations, it is important to prove that the dialogical approach to CTT developed by the two authors in recent papers[11] yields a notion of winning strategy which really corresponds to the notion of CTT demonstration. This is the main aim of the present study. More precisely, we will consider on the one hand Martin Löf's Constructive Type Theory,[12] and, on the other hand, dialogues with play-objects[13] on the other hand. Our purpose is to show the following:

There is a winning **P**-strategy in the dialogical game for φ if and only if there is a CTT demonstration for $p : \varphi$, where p is some proof-object for φ.

One important aspect of the present study is that we restrict ourselves to the logically valid fragment of CTT. The reason is that, once our goal for that fragment is achieved we can extend the result to cover the whole CTT system.

Now, because demonstrations in Martin Löf's CTT are given in the lines of a natural deduction calculus, our demonstration of the equivalence result will be based on Rahman et al. (2009) in which the connection between dialogical semantics and natural deduction has been investigated. However, the present study does not assume knowledge of the papers on dialogues and CTT mentioned afore. There are anyway various significant adjustments that must be made, namely:

[10] In fact there are already some recent applications of the interaction of the dialogical approach and CTT in the context of legal reasoning (Rahman 2014).

[11] Rahman and Clerbout (2013), Rahman et al. (2014).

[12] Cf. Martin-Löf (1984), Nordström et al. (1990), Ranta (1994).

[13] We explain further on the notion of play-objects.

- In the aforementioned paper, the Fitch-style notation has been adopted. In the present study we will not follow that notation and this requires adapting the translation procedure between strategies and CTT demonstrations to a formulation much closer to Gentzen's.
- The paper by Rahman et al. (2009) deals mainly with standard classical propositional logic. Thus, in the context of the present study we need to extend the result to quantifiers and develop an explicit language featuring play-objects for the play level, and proof-objects for the strategical level which yields the CTT notion of proof-object.
- The paper on natural deduction does not provide a systematic method of extracting demonstrations from winning strategies. Such a systematic study has been carried out by Clerbout (2014a, b) in relation to Tableaux. In the present study (one direction of) the method of Clerbout will be adapted to the aims specified above.

This study is thus organized as follows.

Chapters 1 and 2 recall the two frameworks at stake, insisting in particular on the dialogical framework. For one thing, it is true that the dialogical approach is not very widespread in the literature. In addition, the particular system we work with is very recent and thus uncommon even for readers familiar with the dialogical approach. Thus, Chap. 2 contains both the standard account of dialogical games (for readers unfamiliar with it) and the enriched version at stake (for the demonstration of the equivalence result). Our overview of the basics of CTT is briefer mostly because there is a well-established and detailed literature on this approach. We also believe that it is sufficient for the purpose of investigating and showing the connections between the two approaches. However, we refer readers who are not familiar with the CTT framework to detailed introductions such as Martin-Löf (1984), Nordström et al. (1990), and Granström (2011).

Chapter 3 is devoted to the demonstration of the left-to-right direction of the equivalence result.

In Chap. 4 we illustrate the use of the algorithm of Chap. 3 by showing how to transform a specific winning strategy into a CTT demonstration of the Axiom of Choice. The chapter insists on the intensional version of the axiom, but the extensional version is also considered.

Chapter 5 develops the algorithm from CTT demonstrations to dialogical strategies. Some readers may wonder why the material on the Axiom of Choice is placed between the two directions of the demonstration of the equivalence result. The reason is that it relies only on the first direction, covered in Chap. 3, and can thus be understood without the second direction. Actually it could be confusing to place it after proving the second direction in which it plays no part. Hence our choice to place it as an illustration of the procedure developed in Chap. 3, and a kind of interlude before resuming the demonstration of the equivalence result.

We conclude by introducing elements of discussion which are to be developed in subsequent papers. These papers will involve the analysis of the notion of truth rather than the validity and development of a new tableaux system for CTT that follows from the notion of *core* developed in the present study.

Acknowledgments

Many thanks to Göran Sundholm, who inspired the present study during his visit to Lille as visiting professor, and to Johan Georg Granström for patient and thorough discussions on CTT. We are also grateful to Gerhard Heinzmann for his support and for years of inspiring collaboration.

We are thankful to an anonymous reviewer for suggestions that helped us in significantly improving our manuscript and to Zoe Mcconaughey who revised the English and the formal notation. Finally, we are very grateful to the editorial team of the SpringerBriefs series.

The present study is part of an ongoing project in the context of the research-program Argumentation, Decision, Action (ADA) and the ANR11 FRAL 003 01: JURILOG supported by the Maison Européenne des Sciences de l'Homme et de la Société—USR 3185.

Lille
May 2014

Nicolas Clerbout
Shahid Rahman

References

Abramsky, S., and P.A. Mellies. 1999. Concurrent games and full completeness. In *Proceedings of the 14th International Symposium on Logic in Computer Science*, 431–442. Trento: IEEE Computer Society Press.

Austin, J.L. 1946. Other minds. *The Aristotelian Society Supplementary* 20: 148–187.

Blass, A. 1992. A game semantics for linear logic. *Annals of Pure and Applied Logic* 56: 183–220.

Brandom, R. 1994. *Making it Explicit*. Cambridge: Harvard UP.

Brandom, R. 2000. *Articulating Reasons*. Cambridge: Harvard UP.

Clerbout, N. 2014a. First-order dialogical games and tableaux. *Journal of Philosophical Logic* 43 (4): 785–801. [Published first online 2013. doi:10.1007/s10992-013-9289-z].

Clerbout, N. 2014b. *La Sémantique Dialogique. Notions Fondamentales et Éléments de Metathéorie*. London: College Publications.

Dybjer, P. 2012. Tests, games, and Martin-Löf's meaning explanations for intuitionistic type theory. Slides of a talk given at the Congress *Logic and Interactions*, Marseille (France), February 2012.

Ginzburg, J. 2012. *The Interactive Stance: Meaning for Conversation*. Oxford: Oxford University Press.

Girard, J.Y. 1999. On the meaning of logical rules I: Syntax vs. semantics. In *Computational Logic*, ed. U. Berger and H. Schwichtenberg, 215–272. Heidelberg: Springer.

Granström, J. 2011. *Treatise on Intuitionistic Type Theory*. Dordrecht: Springer.

Hodges, W. 2001. Dialogue foundations. A sceptical look. *The Aristotelian Society Supplementary Volume* 75(1): 17–32.

Hodges, W. 2004. Logic and games. In *Stanford Encyclopedia of Philosophy*. http://plato.stanford.edu/entries/logic-dialogical/ (Revised 2013).

Keiff, L. 2007. *Le pluralisme Dialogique: Approches Dynamiques de l'Argumentation Formelle*. PhD thesis, Lille 3.

Lecomte, A. 2011. *Meaning, Logic and Ludics*. London: Imperial College Press.

Lecomte, A., and M. Quatrini. 2010. Pour une étude du langage via l'interaction: Dialogues et sémantique en Ludique. *Mathématiques et Sciences Humaines* 189: 37–67.

Lecomte, A., and M. Quatrini. 2011. Figures of dialogue: A view from ludics. *Synthese* 183(1): 59–85.

Lecomte, A., and S. Tronçon (eds.) 2006–2009. *Ludics, Dialogues and Interaction: PRELUDE Project. Revised Selected Papers*. Berlin: Springer.

Lorenzen, P. 1960. Logik und Agon. In *Atti del XII Congresso Internazionale di Filosofia, Venezia*, 187–194. Sansoni. (Reprinted in Lorenzen and Lorenz, 1978, 1–8).

Lorenzen, P., and K. Lorenz. 1978. *Dialogische Logik*. Darmstadt: Wissenschaftliche Buchgesellschaft.

Marion, M. 2006. Hintikka on Wittgenstein: From language games to game semantics. In *Truth and Games: Essays in Honour of Gabriel Sandu*, volume 78 of *Acta Philosophica Fennica*, 223–242.

Marion, M. 2009. Why play logical games? In *Games: Unifying Logic, Language and Philosophy*, ed. O. Majer, A.V. Pietarinen and T. Tulenheimo, 3–26. Dordrecht: Springer.

Marion, M. 2010. Between saying and doing: From Lorenzen to Brandom and Back. In *Constructions. Essays in Honour of Gerhard Heinzmann*, 489–497. London: College Publications.

Martin-Löf, P. 1984. *Intuitionistic Type Theory*. Naples: Bibliopolis. Notes by Giovanni Sambin of a series of lectures given in Padua, June 1980.

Nordström, B., K. Petersson, and J.M. Smith. 1990. *Programming in Martin-Löf's Type Theory: An Introduction*. Oxford: Oxford University Press.

Prawitz D. 2012. Truth and proof in intuitionism. In *Epistemology versus Ontology. Essays on the Philosophy and Foundations of Mathematics in Honour of Per Martin-Löf*, ed. P. Dybjer, S. Lindström, E. Palmgren, and G. Sundholm, 45–67. Dordrecht: Springer.

Rahman, S., 2014. On hypothetical judgements and Leibniz's notion of conditional right. In *Legal Reasoning and Logic. Past & Present Interactions*, ed. A. Armgardt, P. Canivez and S. Chassagnard-Pinet. Dordrecht: Springer.

Rahman, S., and N. Clerbout. 2013. Constructive type theory and the dialogical approach to meaning. *Baltic International Yearbook of Cognition, Logic and Communication*, vol. 8. doi:10.4148/1944-3676.1077, http://dx.doi.org/10.4148/1944-3676.1077

Rahman, S., N. Clerbout, and L. Keiff. 2009. On dialogues and natural deduction. In *Acts of Knowledge: History, Philosophy and Logic: Essays Dedicated to Göran Sundholm*, ed. G. Primiero and S. Rahman, 301–336. London: College Publications.

Rahman, S., N. Clerbout, and Z. McConaughey. 2014. On play-objects in dialogical games. Towards a dialogical approach to constructive type theory. In *Modestly Radical or Radically Modest. Festschrift for Jean-Paul van Bendegem on the Occasion of his 60th Birthday*, ed. P. Allo and B. van Kerkhove, 127–154. College Publications.

Ranta A. 1988. Propositions as games as types. *Synthese* 76: 377–395.

Ranta, A. 1994. *Type-Theoretical Grammar*. Oxford: Clarendon Press.

Sundholm, G. 1986. Proof-theory and meaning. In *Handbook of Philosophical Logic*, vol. 3, ed. D. Gabbay and F. Guenthner, 471–506. Dordrecht: Reidel.

Sundholm, G. 2001. A plea for logical atavism. In *The Logica Yearbook 2000*, ed. O. Majer, 151–162. Prague: Filosofia.

Sundholm, G. 2013. Inference and consequence in an interpreted language. Talk at the workshop *Proof theory and philosophy*, Groningen.

Contents

List of Figures

List of Tables

Chapter 1
Brief Reminder of Constructive Type Theory

Within Per Martin-Löf's Constructive Type Theory (CTT for short) the logical constants are interpreted through the Curry-Howard correspondence between propositions and sets. A proposition is interpreted as a set whose elements represent the proofs of the proposition. It is also possible to view a set as a problem description in a way similar to Kolmogorov's explanation of the intuitionistic propositional calculus. In particular, a set can be seen as a specification of a programming problem: the elements of the set are then the programs that satisfy the specification (Martin-Löf 1984, p. 7). Furthermore in CTT sets are also understood as types so that propositions can be seen as data (or proof)-types.[1] We will start by a quick overview of the principles of the CTT approach. We will then recall the rules of intuitionistic predicate logic in CTT.

1.1 Fundamentals of the CTT Approach

The general philosophical idea is linked to what has been called the full interpreted[2] approach in which special care is taken, in Martin-Löf's own words (1984, p. 2), to "avoid keeping content and form apart. Instead we will at the same time display certain forms of judgement and inference that are used in mathematical proofs and explain them semantically. Thus, we make explicit what is usually implicitly taken for granted". The *explicitation* task involves bringing into the object language level features that determine meaning and that are usually formulated at the metalevel.

[1]Cf. Nordström et al. (1990) and Granström (2011).

[2]For a thorough discussion on this issue, see Sundholm (1997, 2001).

N. Clerbout and S. Rahman, *Linking Game-Theoretical Approaches with Constructive Type Theory*, SpringerBriefs in Philosophy,
DOI 10.1007/978-3-319-19063-1_1

According to the CTT view, premisses and conclusion of a logical inference are not propositions but judgements.

> A rule of inference is justified by explaining the conclusion on the assumption that the premisses are known. Hence, before a rule of inference can be justified, it must be explained what it is that we must know in order to have the right to make a judgement of any one of the various forms that the premisses and conclusion can have (Martin-Löf 1984, p. 2).

Two further basic tenets of CTT are the following:

1. No entity without type
2. No type without identity

Accordingly, we can take the assertion that an individual is an element of the set A as the assertion that that individual instantiates or exemplifies type A. But what is a type, and how do we differentiate between its examples and those objects that are not of a given type? Or, more fundamentally, what must we know in order to have the right to judge something as a type?

In CTT, a set is defined by specifying its *canonical elements*, the elements that "directly" exemplify the type, and its non-canonical ones, the elements that can be shown, using some prescribed method of transformation, that they are equal (in the type) to a canonical one. Moreover, it is required that the equality between objects of a type be an equivalence relation.[3] This is what the second tenet is about: the introduction of an equivalence relation in a set (an object of the type '*set*').

When we have a type we know, from the semantic explanation of what it means to be a type, what the conditions are for being an object of that type. So if A is a type and we have an object b that satisfies the corresponding conditions, then b is an object of type A, which we formally write $b : A$.[4] Accordingly,

$$b : A \qquad\qquad A\ true$$

can be read as

b is an element of the set A	A has an element
b is a proof of the proposition A	A is true
b fulfils the expectation A	A is fulfilled
b is a solution to the problem A	A has a solution

'*Set*' itself does not instantiate the type *set*, since we do not have a general method for generating all possible ways of building a set. However, given the type '*set*' we can build the objects that instantiate it by the means described above.

[3] For a thorough discussion see Granström (2011, pp. 54–76).

[4] Martin-Löf used the sign "∈" in order to indicate that something, say a, is of some type, say B. He even suggests to understand it as a the copula 'is'. Nordström et al. (1990) also uses this notation while other authors such as Ranta (1994) use the colon. Granström (2011) distinguishes the colon from the epsilon, where the first applies to non-canonical elements and the latter to canonical ones. We will use the colon.

Certain propositional functions can be defined on such a set(-object) once it has been generated. Moreover, the type '*set*' is one of an infinite number of types: there are other types, such as the type *prop*.

Hypotheticals:

The judgements we have introduced so far do not depend on any assumptions. They are categorical judgements. The CTT language has also *hypothetical* judgements of the form

$$B\ type\ (x : A)$$

where A is a type which does not depend on any assumptions, and B is a type when $x : A$ (the *hypothesis* for B). In the case of sets we have b is an element of the set B, under the assumption that x is an element of A:

$$b : B\ (x : A) \qquad \text{(more precisely, } b : el(B)\ (x : el(A)))$$

The explicit introduction of hypotheticals comes with the explicit introduction of appropriate substitution rules. Indeed in the example above, if $a : A$, then the substitution in b of x by a yields an element of B; and if $a = c : A$, then the substitutions in b of x by a and by c are equal elements in B[5]:

$$\frac{a : A \quad b : B\ (x : A)}{b(a/x) : B} \qquad \frac{a = c : A \quad b : B\ (x : A)}{b(a/x) = b(c/x) : B}$$

As pointed out by Granström (2011, p. 112), the form of judgement $b : B\ (x : A)$—or $b : el(B)\ (x : el(A))$—can be generalized in three directions:

(1) Any number of hypotheses will be allowed, not just one;
(2) The set over which a variable ranges may depend on previously introduced variables;
(3) The set B may depend on all the variable introduced.

Such a list of hypotheses will be called a context. Thus we might need the forms of judgement

$b : B\ (\Gamma)$—where Γ is a context (i.e. a list of hypotheses)
$\Gamma : context$

In general, a hypothetical judgement has the form

$$x : A\ (x_1 : A_1\ ,\ x_2 : A_2\ ,\ \ldots\ ,\ x_n : A_n)$$

[5] Cf. Granström (2011, pp. 111–112).

where we already know that A_1 is a type, A_2 is a type in the context $x_1 : A_1$, …, and A_n is a type in the context $x_1 : A_1 , \ x_2 : A_2 , \ \ldots , \ x_{n-1} : A_{n-1}$. That is:

$$\frac{\begin{array}{c} A_1 \ type \\ A_2 \ type \ (x_1 : A_1) \\ A_n \ type \ (x_1 : A_1 , x_2 : A_2 , \ldots , x_{n-1} : A_{n-1}) \\ A \ type \ (x_1 : A_1 , x_2 : A_2 , \ldots , x_n : A_n) \end{array}}{x : A \ (x_1 : A_1 , x_2 : A_2 , \ldots , x_n : A_n)}$$

N.B.: A_1 *type* depends on no hypothesis.

The rules for substitution and equality are generalized accordingly. Hypothetical judgements introduce functions from A to B:

$$f(x) : B \ (x : A)$$

These kinds of expressions can be read in several ways, for example:

$f(x) : B$ *for arbitrary* $x : A$
$f(x) : B$ *under the hypothesis* $x : A$
$f(x) : B$ *provided* $x : A$
$f(x) : B$ *given* $x : A$
$f(x) : B$ *if* $x : A$
$f(x) : B$ *in the context* $x : A$

Propositions are introduced in CTT by laying down what counts as proof of a proposition. Accordingly, a proposition is true if there is such a proof. We write

$$A : prop$$

to formalize the judgement that A is a proposition. Propositional functions are introduced by hypothetical judgements. The hypothetical judgement required to introduce propositional functions is of the form:

$$B(x) : prop \ (x : A)$$

which reads: $B(x)$ is of the type proposition, provided it is applied to elements of the (type-)set A.

We produce propositions from propositional functions by the following rule:

$$\frac{a : A \quad B(x) : prop \ (x : A)}{B(a) : prop}$$

It also requires an appropriate rule defining the equivalence relation within the type *prop*:

$$\frac{a = b : A \quad B(x) : prop \ (x : A)}{B(a) = B(b) : prop}$$

The notion of propositional function as hypothetical judgement allows to (intensionally) introduce subsets by separation:

$$\frac{A : set \quad B(x) : prop \ (x : A)}{\{x : A \mid B(x)\} : set} \qquad \frac{b : A \quad B(b) \ true}{b : \{x : A \mid B(x)\}}$$

This explanation of subsets also justifies the following rules:

$$\frac{b : \{x : A \mid B(x)\}}{b : A} \qquad \frac{b : \{x : A \mid B(x)\}}{B(b) \ true}$$

1.2 The Basic CTT Framework for Intuitionistic Predicate Logic

We now provide a quick presentation of the CTT approach to intuitionistic predicate logic. We start by recalling briefly Martin-Löf's description of the different kinds of rules involved (1984, pp. 24–25).

1.2.1 About the Rules

Since propositions are sets, logical operators are defined as set theoretical operators on sets. Each operator comes with four kinds of rules: the Formation rules, the Introduction rules, the Elimination and the Equality rules.

Formation rules. Including formation rules is one of the most distinctive features of CTT presentations of inferential systems. These rules embody simultaneously the syntax and the explanation of the basic types providing the meaning of the language (involving logical and non-logical constants). They do so by defining the way one can build a set out of other sets or families of sets. Another way of looking at it is to say that the formation rules explain the types of the language.

In fact in the CTT framework every demonstration starts by verifying that the expressions are well-formed or, in other words, that the formation rules are correctly applied. It is in this way that CTT implements the idea of a fully interpreted language. When we read it "bottom-up", a CTT demonstration displays the way that the formation of the expression to be proved rests on the formation of its elements.

The *Introduction rules* for a set give the meaning of the set by specifying what its canonical elements are. *Elimination rules* show how one can define functions on the set defined by the Introduction rules. In short, the Introduction and Elimination rules explain the typing rules for expressions.

Equality rules connect Introduction and Elimination rules by specifying how a function defined by an Elimination rule operates on the canonical elements generated by the Introduction rules.

1.2.2 Intuitionistic Logic in CTT

Our presentation is inspired from Ranta's economy in his 1991 paper.[6] Here are the CTT rules for the set theoretical operators aforementioned and a brief explanation of how the usual logical constants are defined with these operators. For more explanations and details, see for example Martin-Löf (1984, pp. 26–54).

The operator Π (Cartesian product of a family of sets):

$$\frac{A : set \quad \begin{matrix}(x:A)\\ \vdots \\ B(x) : set\end{matrix}}{(\Pi x : A)B(x) : set}\ \Pi F \qquad \frac{\begin{matrix}(x:A)\\ \vdots \\ b(x) : B(x)\end{matrix}}{(\lambda x)b(x) : (\Pi x : A)B(x)}\ \Pi I \qquad \frac{c : (\Pi x : A)B(x) \quad a : A}{Ap(c,a) : B(a)}\ \Pi E$$

$$\frac{a : A \quad \begin{matrix}(x:A)\\ b(x) : B(x)\end{matrix}}{Ap((\lambda x)b(x), a) = b(a) : B(a)}\ \Pi Eq1 \qquad \frac{c : (\Pi x : A)B(x)}{c = (\lambda x)Ap(c,x) : (\Pi x : A)B(x)}\ \Pi Eq2$$

In the Introduction rule, we assume as usual that x does not appear free in any assumption except (those of the form) $x : A$. The two-place function $Ap(x, y)$ is defined by the way it is introduced (in the Elimination rule) and by the way it is computed (in the Equality rules). It can be read as "Application of x to y" and it is a method of obtaining a canonical element of $B(a)$. See Martin-Löf (1984, pp. 28–29).

Universal quantification and material implication are then defined as follows:

$(\forall x : A)B(x) = (\Pi x : A)B(x) : prop$ for $A : set$ and $B(x) : prop\ (x : A)$
$A \to B = (\Pi x : A)B : prop$ for $A : prop$ and $B : prop$

The operator Σ (disjoint union of a family of sets):

$$\frac{A : set \quad \begin{matrix}(x:A)\\ \vdots \\ B(x) : set\end{matrix}}{(\Sigma x : A)B(x) : set}\ \Sigma F \qquad \frac{a : A \quad b : B(a)}{(a,b) : (\Sigma x : A)B(x)}\ \Sigma I$$

$$\frac{c : (\Sigma x : A)B(x) \quad \begin{matrix}(x : A, y : B(x))\\ \vdots \\ d(x,y) : C((x,y))\end{matrix}}{E(c,(x,y)d(x,y)) : C(c)}\ \Sigma E \qquad \frac{a : A \quad b : B(a) \quad \begin{matrix}(x : A, y : B(x))\\ d(x,y) : C((x,y))\end{matrix}}{E(a,b,(x,y)d(x,y)) = d(a,b) : C((a,b))}\ \Sigma Eq$$

In the Elimination rule, $E(c, (x, y)d(x, y))$ stands for "Execute c. It yields a pair: substitute it for (x, y) in $d(x, y)$". See Martin-Löf (1984, p. 40).

[6] Cf. Ranta (1991, Sect. 3).

Existential quantification and conjunction are then defined as follows:

$$(\exists x : A)B(x) = (\Sigma x : A)B(x) : prop \quad \text{for } A : set \text{ and } B(x) : prop\ (x : A)$$
$$A \wedge B = (\Sigma x : A)B : prop \quad \text{for } A : prop \text{ and } B : prop$$

Remark In the case of conjunction, we obtain the two standard Elimination rules by simply choosing C to be A or B and by defining the left and right projections $p(c)$ and $q(c)$ as follows: $p(c) \equiv E(c, (x, y)x)$ and $q(c) \equiv E(c, (x, y)y)$. We then obtain the rules

$$\frac{c : A \wedge B}{p(c) : A} \wedge E1 \qquad \frac{A \wedge B}{q(c) : B} \wedge E2$$

See Martin-Löf (1984, pp. 44–46).

The operator + (disjoint union or coproduct of two sets):

$$\frac{A : set \quad B : set}{A + B : set} +F \qquad \frac{a : A}{i(a) : A + B} +I1 \qquad \frac{b : B}{j(b) : A + B} +I2$$

$$\frac{c : A + B \quad \begin{matrix}(x : A)\\ \vdots \\ d(x) : C(i(x))\end{matrix} \quad \begin{matrix}(y : B)\\ \vdots \\ e(y) : C(j(y))\end{matrix}}{D(c, (x)d(x), (y)e(y)) : C(c)} +E \qquad \frac{a : A \quad \begin{matrix}(x : A)\\ \vdots \\ d(x) : C(i(x))\end{matrix} \quad \begin{matrix}(y : B)\\ \vdots \\ e(y) : C(j(y))\end{matrix}}{D(i(b), (x)d(x), (y)e(y)) = d(a) : C(i(a))} +eq1$$

$$\frac{b : B \quad \begin{matrix}(x : A)\\ \vdots \\ d(x) : C(i(x))\end{matrix} \quad \begin{matrix}(y : B)\\ \vdots \\ e(y) : C(j(y))\end{matrix}}{D(j(b), (x)d(x), (y)e(y)) = e(b) : C(j(b))} +eq2$$

where i and j are two new primitive constants giving the information that an element of $A + B$ comes from A or B, and which one of the two is the case. In the Elimination rule, $D(c, (x)d(x), (y)e(y))$ reads: "Execute c. If it yields the canonical element $i(a)$ then substitute a for x in $d(x)$; if it yields $j(b)$ then substitute b for y in $e(y)$."

Disjunction is then simply defined as follows:

$$A \vee B = A + B : prop \quad \text{for } A : prop \text{ and } B : prop$$

The absurdum (or "bottom") $\bot$:

$$\frac{}{\bot : set} \bot F \qquad \frac{x : \bot}{a : A} \bot E$$

for any x.

The symbol $\bot$ is merely another name for the empty set, and since it is always a set, the formation rule acknowledges that $\bot$ is always a proposition. There is no (need for an) Introduction rule for $\bot$. It is introduced only by means of the Elimination rule for material implication (for expressions of the form $A \rightarrow \bot$). This is in accordance with the standard view of $\neg A$ as an abbreviation for $A \rightarrow \bot$ in intuitionistic logic. Since there is no Introduction rule for $\bot$, there cannot be an equality rule relating the Elimination rule to it. One last remark: the Elimination rule expresses the so-called *ex falso sequitur quodlibet*: the judgement that some x is of type $\bot$ is contradictory since $\bot$ is nothing but the empty set. Hence from it one may conclude $a : A$. Notice however that these remarks are somehow simplistic: in CTT the rules for $\bot$ are actually derived from the rules for finite sets, as explained in Martin-Löf (1984, pp. 65–67).

References

Granström, J. 2011. *Treatise on Intuitionistic Type Theory*. Dordrecht: Springer.

Martin-Löf, P. 1984. *Intuitionistic Type Theory*. Naples: Bibliopolis. Notes by Giovanni Sambin of a series of lectures given in Padua, June 1980.

Nordström, B., K. Petersson, and J.M. Smith. 1990. *Programming in Martin-Löf's Type Theory: An Introduction*. Oxford: Oxford University Press.

Ranta, A. 1991. Constructing possible worlds. *Theoria* 57(1–2): 77–99.

Ranta, A. 1994. *Type-Theoretical Grammar*. Oxford: Clarendon Press.

Sundholm, G. 1997. Implicit epistemic aspects of constructive logic. *Journal of Logic, Language and Information* 6(2): 191–212.

Sundholm, G. 2001. A plea for logical atavism. In *The Logica Yearbook 2000*, ed. O. Majer, 151–162. Prague: Filosofia.

Chapter 2
Dialogues with Play-Objects

The dialogical approach to logic is not a specific logical system but rather a rule-based semantic framework in which different logics can be developed, combined and compared. An important point is that there are different kinds of rules fixing meaning. This feature of the underlying semantics of the dialogical approach has often caused it to be called a *pragmatist* semantics.[1]

More clearly, in a dialogue two parties argue about a thesis while respecting certain fixed rules. The player stating the thesis is called Proponent (**P**) and his rival, the one contesting the thesis, is called Opponent (**O**). In their original form, dialogues were designed in such a way that each play ends after a finite number of moves with one player winning, and the other losing. Actions or moves in a dialogue are often understood as speech-acts involving declarative utterances (i.e. posits) and interrogative utterances (i.e. requests). The point is that the rules of the dialogue do not operate on expressions or sentences isolated from the act of uttering them. The rules are divided into particle rules or rules for logical constants (*Partikelregeln*) and

[1]The main original papers are collected in Lorenzen and Lorenz (1978). For a historical overview see Lorenz (2001). For a presentation about the initial role of the framework as a foundation for intuitionistic logic, see Felscher (1985). Other papers have been collected more recently in Lorenz (2008, 2010a, b). A detailed account of recent developments, since Rahman (1993), can be found in Rahman and Keiff (2005), Keiff (2009) and Rahman (2012). For the underlying metalogic see Clerbout (2014a, c). For textbook presentations: Kamlah and Lorenzen (1972, 1984), Lorenzen and Schwemmer (1975), Redmond and Fontaine (2011) and Rückert (2011a). For the key role of dialogic in linking dialectics and logic again, see Rahman and Keiff (2010). Keiff (2004a, b, 2007) and Rahman (2009) study the dialogical approach to Modal Logic. Fiutek et al. (2010) study the dialogical approach to belief revision. Clerbout et al. (2011) study Jain Logic in the dialogical framework. Popek (2012) developed a dialogical reconstruction of medieval obligationes. Rahman and Tulenheimo (2009) study the links between GTS and Dialogical games. For other books see Redmond (2010)— on fiction and dialogic—Fontaine (2013)—on intentionality, fiction and dialogues—and Magnier (2013)—on dynamic epistemic logic (van Ditmarsch et al. 2007) and legal reasoning in a dialogical framework.

N. Clerbout and S. Rahman, *Linking Game-Theoretical Approaches with Constructive Type Theory*, SpringerBriefs in Philosophy,
DOI 10.1007/978-3-319-19063-1_2

structural rules (*Rahmenregeln*). The structural rules establish the general course of a dialogue game, whereas the particle rules regulate the moves (or utterances) that are either requests (against the rival's moves) or answers (to the requests).

The following points are crucial to the dialogical approach[2]:

1. There is a distinction between the local meaning (rules for logical constants) and the global meaning (included in the structural rules that determine how to play);
2. The local meaning is player-independent;
3. There is a distinction between the play level (local winning or winning of a play) and the strategic level (existence of a winning strategy);
4. The notion of validity amounts to the existence of a winning strategy independently of any model instead of the existence of a winning strategy for every model;
5. Non formal and formal plays are differentiated. Formal plays concern plays in which positing elementary sentences does not require a meta-language level providing their truth.

In the framework of Constructive Type Theory propositions are sets whose elements are called proof-objects. When such a set is not empty, then the proposition has a proof and it is true. In his 1988 paper, Ranta proposed a way to use this approach in relation to game-theoretical approaches. Ranta took Hintikka's Game Theoretical Semantics as a case study, but his point does not depend on that particular framework. Ranta's idea was that in the context of game-based approaches, a proposition is a set of winning strategies for the player positing the proposition.[3] In game-based approaches, the notion of truth is found at the level of such winning strategies. Ranta's idea should therefore let us apply safely and directly methods taken from Constructive Type Theory to cases of game-based approaches.

But from the perspective of game theoretical approaches, reducing a game to a set of winning strategies is quite unsatisfactory, especially when it comes to a theory of meaning. This is particularly clear in the dialogical approach in which different levels of meaning are carefully distinguished. There is thus the level of strategies which is a level of meaning analysis, but there is also a level prior to it which is usually called the level of plays. The role of the latter level for developing an analysis is crucial according to the dialogical approach, as pointed out by Kuno Lorenz in his 2001 paper, p. 258:

> [...] for an entity [A] to be a proposition there must exist a dialogue game associated with this entity [...] such that an individual play of the game where A occupies the initial position [...] reaches a final position with either win or loss after a finite number of moves [...]

For this reason we would rather have propositions interpreted as sets of what we shall call play-objects and read an expression

$$p : \varphi$$

as "p is a play-object for φ".

---

[2]Cf. Rahman (2012).

[3]That player can be called Player 1, Myself, or Proponent.

Thus, Ranta's work on proof-objects and strategies constitutes the end, not the beginning, of the dialogical project.

In order to present the dialogical framework which we will link with CTT, we will proceed in two steps. In the first section of this chapter, we present quite briefly the standard dialogical framework. The purpose is to introduce or recall the basics of the dialogical approach before delving into the more sophisticated system which we are interested in. This progressive introduction should be particularly helpful to readers not familiar with the dialogical approach. Covering the basics in the first section allows us to focus in the rest of the chapter on the framework modifications triggered by adding play-objects to the language.

2.1 Standard Dialogical Games

Let $\mathscr{L}$ be a first-order language built as usual upon the propositional connectives, the quantifiers, a denumerable set of individual variables, a denumerable set of individual constants and a denumerable set of predicate symbols (each with a fixed arity).

We extend the language $\mathscr{L}$ with two labels **O** and **P**, standing for the players of the game, and the two symbols '!' and '?'. When the identity of the player does not matter, we use variables **X** or **Y** (with $\mathbf{X} \neq \mathbf{Y}$). A move is an expression of the form '**X**-e', where e is either of the form $!\varphi$ for some sentence φ of $\mathscr{L}$ or of the form '$?[\varphi_1, \ldots, \varphi_n]$'.

The particle (or local) rules for standard dialogical games are given in Table 2.1.

Table 2.1 Particle rules in standard dialogical games

Posit	Challenge	Defence
$\mathbf{X}\ !\varphi \wedge \psi$	$\mathbf{Y}\ ?[\varphi]$ or $\mathbf{Y}\ ?[\psi]$	$\mathbf{X}\ !\varphi$ / $\mathbf{X}\ !\psi$
$\mathbf{X}\ !\varphi \vee \psi$	$\mathbf{Y}\ ?[\varphi, \psi]$	$\mathbf{X}\ !\varphi$ or $\mathbf{X}\ !\psi$
$\mathbf{X}\ !\varphi \rightarrow \psi$	$\mathbf{Y}\ !\varphi$	$\mathbf{X}\ !\psi$
$\mathbf{X}\ !\neg\varphi$	$\mathbf{Y}\ !\varphi$	–
$\mathbf{X}\ !\forall x\varphi$	$\mathbf{Y}\ ?[\varphi(x/a_i)]$	$\mathbf{X}\ !\varphi(x/a_i)$
$\mathbf{X}\ !\exists x\varphi$	$\mathbf{Y}\ ?[\varphi(x/a_1), \ldots, \varphi(x/a_n)]$	$\mathbf{X}\ !\varphi(x/a_i)$ with $1 \leq i \leq n$

In this table, the a_is are individual constants and $\varphi(x/a_i)$ denotes the formula obtained by replacing every free occurrence of x in φ by a_i. When a move consists in a question of the form '$?[\varphi_1, \ldots, \varphi_n]$', the other player chooses one formula among $\varphi_1, \ldots, \varphi_n$ and plays it. We thus distinguish conjunction from disjunction and universal quantification from existential quantification in terms of which player has a choice. With conjunction and universal quantification, the challenger chooses which formula he asks for. With disjunction and existential quantification, it is the defender who can choose between various formulas. Notice that there is no defence in the particle rule for negation.

Particle rules provide an abstract description of how the game can proceed locally: they specify the way a formula can be challenged and defended according to its main logical constant. In this way the particle rules govern the local level of meaning. Strictly speaking, the expressions occurring in the table above are not actual moves because they feature formula schemata and the players are not specified. Moreover, these rules are indifferent to any particular situation that might occur during the game. For these reasons we say that the description provided by the particle rules is abstract.

Since the players' identities are not specified in these rules, particle rules are symmetric: the rules are the same for the two players. The local meaning being symmetric (in this sense) is one of the greatest strengths of the dialogical approach to meaning. It is in particular the reason why the dialogical approach is immune to a wide range of trivializing connectives such as Prior's *tonk*.[4]

The expressions occurring in particle rules are all move schematas. The words "challenge" and "defence" are convenient to name certain moves according to their relation with other moves which can be defined in the following way. Let σ be a sequence of moves. The function p_σ assigns a position to each move in σ, starting with 0. The function F_σ assigns a pair $[m, Z]$ to certain moves N in σ, where m denotes a position smaller than $p_\sigma(N)$ and Z is either C or D, standing respectively for "challenge" and "defence". That is, the function F_σ keeps track of the relations of challenge and defence as they are given by the particle rules. Consider for example the following sequence σ:

$$\mathbf{P}\,!\varphi \wedge \psi, \mathbf{P}\,!\chi \wedge \psi, \mathbf{O}\,?[\varphi], \mathbf{P}\,!\varphi$$

In this sequence we have for example $p_\sigma(\mathbf{P}\,!\chi \wedge \psi) = 1$.

A *play* is a legal sequence of moves, i.e., a sequence of moves which observes the game rules. Particle rules are not the only rules which must be observed in this respect. In fact, it can be said that the second kind of rules named *structural rules* are the ones giving the precise conditions under which a given sequence is a play. The dialogical game for φ, written $\mathscr{D}(\varphi)$, is the set of all plays with φ being the *thesis* (see the Starting Rule below). The structural rules are the following:

SR0 (Starting Rule). Let φ be a complex sentence of $\mathscr{L}$ and i, j be positive integers. For every $\varsigma \in \mathscr{D}(\varphi)$ we have:

[4] See Rahman et al. (2009) and Rahman (2012).

- $p_\varsigma(\mathbf{P}\,!\varphi) = 0$,
- $p_\varsigma(\mathbf{O}\,\texttt{n} := \texttt{i}) = 1$,
- $p_\varsigma(\mathbf{P}\,\texttt{m} := \texttt{j}) = 2$.

In other words, any play ς in $\mathscr{D}(\varphi)$ starts with **P** positing φ. We call φ the thesis of both the play and the dialogical game. After that, the Opponent and the Proponent successively choose a positive integer called repetition rank. The role of these integers is to ensure that every play ends after finitely many moves in the way specified by the next structural rule.

SR1 (Classical game-playing Rule).

- Let $\varsigma \in \mathscr{D}(\varphi)$. For every M in ς with $p_\varsigma(M) > 2$ we have $F_\varsigma(M) = [m', Z]$ with $m' < p_\varsigma(M)$ and $Z \in \{C, D\}$.
- Let $\texttt{r}$ be the repetition rank of player **X** and $\varsigma \in \mathscr{D}(\varphi)$ such that
  - the last member of ς is a **Y**-move,
  - M_0 is a **Y**-move of position m_0 in ς,
  - $M_1, \ldots, M_n$ are **X**-moves in ς such that $F_\varsigma(M_1) = \cdots = F_\varsigma(M_n) = [m_0, Z]$.

  Consider the sequence[5] $\varsigma' = \varsigma \star N$ where N is an **X**-move such that $F_{\varsigma'}(N) = [m_0, Z]$. We have $\varsigma' \in \mathscr{D}(\varphi)$ only if $n < \texttt{r}$.

The first part of the rule states that, after repetition ranks have been chosen, every move is either a challenge or a defence. The second part ensures finiteness of plays by setting the player's repetition rank as the maximum number of times he can challenge or defend against a given move by the other player.

SR2 (Formal Rule). Let ψ be an elementary sentence, N be the move $\mathbf{P}\,!\psi$ and M be the move $\mathbf{O}\,!\psi$. A sequence ς of moves is a play only if we have: if $N \in \varsigma$ then $M \in \varsigma$ and $p_\varsigma(M) < p_\varsigma(N)$.

That is, the Proponent can play an elementary sentence only if the Opponent has played it previously. The Formal Rule is one of the characteristic features of the dialogical approach: other game-based approaches do not have it.

Helge Rückert pointed out that the Formal Rule triggers a novel notion of validity: *Geltung* (Legitimacy).[6] Indeed with this rule the dialogical framework comes with an internal account for elementary sentences: an account in terms of interaction only, without depending on metalogical meaning explanations for the non-logical vocabulary. More prominently this means that the dialogical account does not rely—contrary to Hintikka's GTS games—on the model-theoretical approach to meaning for atomic formulas.

From there Rückert claims, and on this point we disagree with him, that *Geltung* is the idea that interaction emerges without knowing (or without needing to know) what the meaning of elementary sentences are. We disagree because the question of the meaning of elementary sentences (and more generally, of non-logical vocabulary)

---

[5] We use $\varsigma \star N$ to denote the sequence obtained by adding move N to the play ς.

[6] See Rückert (2011b).

cannot be disregarded if the dialogical framework is meant to provide a general theory of meaning.

In our view, Rückert's interpretation of *Geltung* unfortunately dissolves the question of the meaning of elementary sentences in the Formal Rule. This is mainly due to the fact that the standard version of the framework does not have the means to express a semantic at the object language level in terms of asking and giving reasons for elementary sentences. As a consequence, the standard formulation simply relies on the Formal Rule which amounts to entitle **P** to copy-cat the elementary sentences brought forward by **O**. According to us, the introduction of play-objects provides a solution to this without giving up the internal aspect linked with *Geltung*. We will develop this idea when giving the particle rules in Sect. 2.3 and after having introduced a "Modified Formal Rule" in Sect. 2.4.

Here is some terminology for the last structural rule in standard dialogical games. A play is called *terminal* when it cannot be extended by further moves in compliance with the rules. We say it is **X**-terminal when the last move in the play is an **X**-move.

SR3 (Winning Rule). Player **X** wins the play ς only if it is **X**-terminal.

Consider for example the following sequences of moves:

$$\mathbf{P}\,!Q(a) \wedge Q(b), \mathbf{O}\,\mathtt{n} := 1, \mathbf{P}\,\mathtt{m} := 6, \mathbf{O}\,?[Q(a)], \mathbf{P}\,!Q(a)$$
$$\mathbf{P}\,!Q(a) \to Q(a), \mathbf{O}\,\mathtt{n} := 1, \mathbf{P}\,\mathtt{m} := 12, \mathbf{O}\,!Q(a), \mathbf{P}\,!Q(a)$$

The first one is not a play because it breaks the Formal Rule: with his last move, the Proponent plays an elementary sentence which the Opponent has not played beforehand. By contrast, the second sequence is a play in $\mathscr{D}(Q(a) \to Q(a))$.

We often use a convenient table notation for plays. For example, we can write this play as follows:

	O			P	
				$!\,Q(a) \to Q(a)$	0
1	n := 1			m := 12	2
3	$!\,Q(a)$	(0)		$!\,Q(a)$	4

The numbers in the external columns are the positions of the moves in the play. When a move is a challenge, the position of the challenged move is indicated in the internal columns, as with move 3 in this example. Notice that such tables carry the information given by the functions p and F in addition to represent the play.

However, when we want to consider several plays together—for example when building a strategy—such tables are not that perspicuous. So we do not use them to deal with dialogical games for which we prefer another perspective. The *extensive form* of the dialogical game $\mathscr{D}(\varphi)$ is simply the tree representation of it, also often called the game-tree. More precisely, the extensive form $\mathfrak{E}_\varphi$ of $\mathscr{D}(\varphi)$ is the tree (T, ℓ, S) such that:

(i) Every node t in T is labelled with a move occurring in $\mathscr{D}(\varphi)$.
(ii) $\ell : T \mapsto \mathbb{N}$.
(iii) $S \subseteq T^2$ with the following:

- There is a unique t_0 (the root) in T such that $\ell(t_0) = 0$, and t_0 is labelled with the thesis of the game.
- For every $t \neq t_0$ there is a unique t' such that $t'St$.
- For every t and t' in T, if tSt' then $\ell(t') = \ell(t) + 1$.
- Let $\varsigma \in \mathcal{D}(\varphi)$ such that $p_\varsigma(M') = p_\varsigma(M) + 1$. If t and t' are respectively labelled with M and M', then tSt'.

Many metalogical results about dialogical games are obtained by leaving the level of rules and plays to move to the level of strategies. Significant among these results are the ones concerning the existence of winning strategies for a player. We will now define these notions and give examples of such results.

A *strategy* for player **X** in $\mathcal{D}(\varphi)$ is a function which assigns an **X**-move M to every non terminal play ς having a **Y**-move as last member such that extending ς with M results in a play. An **X**-strategy is winning if playing according to it leads to **X**'s victory no matter how **Y** plays.

Strategies can be considered from the perspective of extensive forms: the extensive form of an **X**-strategy s in $\mathcal{D}(\varphi)$ is the tree-fragment $\mathfrak{S}_\varphi = (T_s, \ell_s, S_s)$ of $\mathfrak{E}_\varphi$ such that:

(i) The root of $\mathfrak{S}_\varphi$ is the root of $\mathfrak{E}_\varphi$,
(ii) Given a node t in $\mathfrak{E}_\varphi$ labelled with an **X**-move, we have $t' \in T_s$ and tS_st' whenever tSt'.
(iii) Given a node t in $\mathfrak{E}_\varphi$ labelled with a **Y**-move and with at least one t' such that tSt', there is a unique t_s in T_s with tS_st_s and t_s is labelled with the **X**-move prescribed by s.

Here are some results pertaining to the level of strategies.[7]

- Winning **P**-Strategies and Leaves. *Let w be a winning* **P***-strategy in $\mathcal{D}(\varphi)$. Then every leaf in the extensive form $\mathfrak{W}_\varphi$ of w is labelled with a* **P** *elementary sentence.*
- Determinacy. *There is a winning* **X***-strategy in $\mathcal{D}(\varphi)$ if and only if there is no winning* **Y***-strategy in $\mathcal{D}(\varphi)$.*
- Soundness and Completeness of Tableaux. *Consider first-order tableaux and first-order dialogical games. There is a tableau proof for φ if and only if there is a winning* **P***-strategy in $\mathcal{D}(\varphi)$.*

  The fact that the existence of a winning **P**-strategy coincides with validity (there is a winning **P**-strategy in $\mathcal{D}(\varphi)$ if and only if φ is valid) follows from the soundness and completeness of the tableau method with respect to model-theoretical semantics.

Regarding several results, extensive forms of strategies have key parts: one of the parts of a winning strategy, called the *core* of the strategy, is actually that on which one works when considering translation algorithms such as the procedures given in Chaps. 3 and 5. We will give the details in Chap. 3, but the basic idea behind the notion of core is to get rid of redundant information (for example, different orders of

[7]These results are proven, together with others, in Clerbout (2014c).

moves) which we find in extensive forms of strategies. Now that we have recalled the standard dialogical approach, we will focus on the enriched dialogical framework we are interested in for the equivalence result with CTT.

2.2 The Formation of Propositions

Before delving into the details about play-objects, let us first discuss the issue of forming expressions and especially propositions in the dialogical approach.

It is presupposed in standard dialogical systems that the players use well-formed formulas (wff). The well formation can be checked at will, but only with the usual meta reasoning by which the formula is checked to indeed observe the definition of a wff. The first enrichment we want to make is to allow players to question the status of expressions, and in particular to ask if a certain expression is a proposition. We thus start with rules explaining dialogically the *formation* of propositions. These rules are local rules which are added to the particle rules giving the local meaning of logical constants (see next section).

A remark before displaying the formation rules: because the dialogical theory of meaning is based on argumentative interaction, dialogues feature expressions which are not posits of sentences. That is they also feature requests, used for challenges, as the formation rules below and the particle rules in the next section illustrate. Because of the *no entity without type* principle, it seems at first glance that we should specify the type of these actions during a dialogue: the type "*formation-request*". It turns out we should not: an expression such as "$?_F$: *formation-request*" is a judgement that some action $?_F$ is a formation-request, which should not be confused with the actual act of requesting. We also consider that the force symbol $?_F$ makes the type explicit. Hence the way requests are written in rules and dialogues in this work.

The formation rules are given in Table 2.2, p. 17. Notice that a posit '$\bot : prop$' cannot be challenged: this is the dialogical account of the fact that the falsum $\bot$ is by definition a proposition.

Table 2.2 Formation rules

Posit	Challenge (when different challenges are possible, the challenger chooses)	Defence
$\mathbf{X}\,!\,\Gamma : set$	$\mathbf{Y}\ ?_{can}\ \Gamma$ or $\mathbf{Y}\ ?_{gen}\ \Gamma$ or $\mathbf{Y}\ ?_{eq}\ \Gamma$	$\mathbf{X}\,!\,a_1 : \Gamma\,, \mathbf{X}\,!\,a_2 : \Gamma\,, \ldots$ **X** gives the canonical elements of Γ $\mathbf{X}\,!\,a_i : \Gamma \Rightarrow a_j : \Gamma$ **X** provides a generation method for Γ **X** gives the equality rule for Γ (see Sect. 2.3)
$\mathbf{X}\,!\,\varphi \vee \psi : prop$	$\mathbf{Y}\ ?_{F_{\vee 1}}$ or $\mathbf{Y}\ ?_{F_{\vee 2}}$	$\mathbf{X}\,!\,\varphi : prop$ respectively $\mathbf{X}\,!\,\psi : prop$
$\mathbf{X}\,!\,\varphi \wedge \psi : prop$	$\mathbf{Y}\ ?_{F_{\wedge 1}}$ or $\mathbf{Y}\ ?_{F_{\wedge 2}}$	$\mathbf{X}\,!\,\varphi : prop$ respectively $\mathbf{X}\,!\,\psi : prop$
$\mathbf{X}\,!\,\varphi \rightarrow \psi : prop$	$\mathbf{Y}\ ?_{F_{\rightarrow 1}}$ or $\mathbf{Y}\ ?_{F_{\rightarrow 2}}$	$\mathbf{X}\,!\,\varphi : prop$ respectively $\mathbf{X}\,!\,\psi : prop$
$\mathbf{X}\,!\,(\forall x : A)\varphi(x) : prop$	$\mathbf{Y}\ ?_{F_{\forall 1}}$ or $\mathbf{Y}\ ?_{F_{\forall 2}}$	$\mathbf{X}\,!\,A : set$ respectively $\mathbf{X}\,!\,\varphi(x) : prop\ (x : A)$
$\mathbf{X}\,!\,(\exists x : A)\varphi(x) : prop$	$\mathbf{Y}\ ?_{F_{\exists 1}}$ or $\mathbf{Y}\ ?_{F_{\exists 2}}$	$\mathbf{X}\,!\,A : set$ respectively $\mathbf{X}\,!\,\varphi(x) : prop\ (x : A)$
$\mathbf{X}\,!\,B(k) : prop$ for atomic B	$\mathbf{Y}\ ?_{F}$	$\mathbf{X}\ sic(n)$ **X** indicates that **Y** posited it at move n
$\mathbf{X}\,!\,\bot : prop$	–	–

The next rule is not a formation rule per se but rather a substitution rule.[8] When φ is an elementary sentence, the substitution rule helps explaining the formation of such sentences.

Posit-substitution
When a list of variables occurs in a posit with proviso, the challenger **Y** can ask **X** to replace those variables: he does so by positing an instantiation of the proviso, in which he (**Y**) is the one who chooses the instantiations for the variables.[9]

[8]It is an application of the original rule from CTT given in Ranta (1994, p. 30).

[9]More precisely: in the case where the defender did not commit himself to the proviso. The dialogical approach allows a distinction here which we discuss in the next section.

Posit	Challenge	Defence
$\mathbf{X} ! \pi(x_1, \ldots, x_n)$ $(x_i : A_i)$	$\mathbf{Y} ! \tau_1 : A_1, \ldots, \tau_n : A_n$	$\mathbf{X} ! \pi(\tau_1, \ldots, \tau_n)$

A particular case of posit substitution is when the challenger simply posits the whole assumption as it is without introducing new instanciation terms. This is particularly useful in the case of formation plays: see an application in the example given in Table 2.4, p. 20.

Posit	Challenge	Defence
$\mathbf{X} ! \pi(\tau_1, \ldots, \tau_n)$ $(\tau_i : A_i)$	$\mathbf{Y} ! \tau_1 : A_1, \ldots, \tau_n : A_n$	$\mathbf{X} ! \pi(\tau_1, \ldots, \tau_n)$

Remarks on the formation dialogues:

(a) Conditional formation posits: A crucial feature of formation rules is that they enable the displaying of the syntactic and semantic presuppositions of a given thesis which can thus be examined by the Opponent before running the actual dialogue on the thesis. For instance if the thesis amounts to positing φ, then the Opponent can ask for its formation before launching an attack. Defending on the formation of φ might bring the Proponent to posit that φ is a proposition, provided for instance that A being a set is conceded. In this situation the Opponent might concede A is a set, but only after the Proponent displayed the constitution of A.

(b) Elementary sentences, definitional consistency and material-analytic dialogues: Following the idea of formation rules through and through, the defence $sic(n)$ for elementary sentences is somehow unsatisfactory as it does not really explore the formation of the expression. A defence which applies fitting predicator rules previously conceded, if such a concession has been made, would be a possibility. See Rahman and Clerbout (2014). What would then happen is that the challenge of elementary sentences would be based on the definitional consistency of the use of the conceded predicator rules. This is what we think material dialogues are about: definitional consistency dialogues. This leads to the following material analytic rule for formation dialogues:

> **O**'s elementary sentences cannot be challenged, however **O** can challenge an elementary sentence (posited by **P**) iff she herself (the Opponent) did not posit it before.

Remark Once **P** forced **O** to concede the elementary sentence in the formation dialogue, the dialogue proceeds using the copy-cat strategy. The version of the rule we work with, in which the defence is $sic(n)$, is related to that.

By way of illustration, Table 2.3 gives an example where the Proponent posits the thesis $(\forall x : A)(B(x) \to C(x)) : prop$ given that $A : set$, $B(x) : prop$ $(x : A)$ and $C(x) : prop$ $(x : A)$. The three provisos appear as initial concessions by the Opponent.[10] Normally we should give all the rules of the game before giving an example, but we make an exception here because the standard structural rules of Sect. 2.1 are enough to understand the following plays. We can thus focus on illustrating the way formation rules can be used.

[10]The example comes from Ranta (1994, p. 31).

Table 2.3 Example of a formation dialogue 1

	O			P	
I	$!A : set$				
II	$!B(x) : prop\ (x : A)$				
III	$!C(x) : prop\ (x : A)$				
				$!(\forall x : A)(B(x) \to C(x)) : prop$	0
1	$\mathtt{n} := 1$			$\mathtt{m} := 2$	2
3	$?_{F\forall 1}$	(0)		$!A : set$	4

Explanations:

- I to III: **O** concedes that A is a set and that $B(x)$ and $C(x)$ are propositions provided x is an element of A.
- Move 0: **P** posits that the main sentence, universally quantified, is a proposition (under the concessions made by **O**).
- Moves 1 and 2: the players choose their repetition ranks.[11]
- Move 3: **O** challenges the thesis by asking the left-hand part as specified by the formation rule for universal quantification.
- Move 4: **P** responds by positing that A is a set. This has already been granted with the premiss I so even if **O** were to challenge this posit, the Proponent could refer to this initial concession. Later, we will introduce the structural rule $SR3$ to deal with this phenomenon (see Sect. 2.4). Thus **O** has no further possible move, the dialogue ends here and is won by **P**.[12]

Obviously, this dialogue does not cover all the aspects related to the formation of $(\forall x : A)(B(x) \to C(x))$. Notice however that the formation rules allow an alternative move for the Opponent's move 3.[13] Hence another possible course of action for **O** arises (Table 2.4).

[11]The device of repetition ranks is introduced in the structural rules which we present in Sect. 2.4. See also Clerbout (2014a, b, c) for detailed explanations on this notion.

[12]See Sect. 2.4.

[13]As a matter of fact, increasing her repetition rank would allow her to play the two alternatives for move 3 within a single play. But increasing the Opponent's rank usually yields redundancies (Clerbout 2014b, c) making things harder to understand for readers not familiar with the dialogical approach. Hence our choice to divide the example into different simple plays.

Table 2.4 Example of a formation dialogue 2

	O			P	
I	$!\, A : set$				
II	$!\, B(x) : prop\ (x : A)$				
III	$!\, C(x) : prop\ (x : A)$				
				$!\, (\forall x : A)(B(x) \to C(x)) : prop$	0
1	n := 1			m := 2	2
3	$?_{F\forall 2}$	(0)		$!\, B(x) \to C(x) : prop\ (x : A)$	4
5	$!\, x : A$	(4)		$!\, B(x) \to C(x) : prop$	6
7	$?_{F\to 1}$	(6)		$!\, B(x) : prop$	10
9	$!\, B(x) : prop$		(*II*)	$!\, x : A$	8

Explanations:

The second dialogue starts like the first one until move 2. Then:

- Move 3: This time **O** challenges the thesis by asking for the right-hand part.
- Move 4: **P** responds, positing that $B(x) \to C(x)$ is a proposition provided $x : A$.
- Move 5: **O** uses the substitution rule to challenge move 4 by granting the proviso.
- Move 6: **P** responds by positing that $B(x) \to C(x)$ is a proposition.
- Move 7: **O** then challenges move 6 by asking the left-hand part as specified by the formation rule for material implication.

To defend this **P** needs to make an elementary move. But since **O** has not played it yet, **P** cannot defend it at this point. Thus:

- Move 8: **P** launches a counterattack against assumption II by applying the substitution rule.
- Move 9: **O** answers to move 8 and posits that $B(x)$ is a proposition.
- Move 10: **P** can now defend in reaction to move 7 and win this dialogue.

Then again, there is another possible path for the Opponent because she has another possible choice for her move 7, namely asking the right-hand part. This yields a dialogue similar to the one above except that the last moves are about $C(x)$ instead of $B(x)$.

By displaying these various possibilities for the Opponent, we have entered the *strategical* level. This is the level at which the question of the good formation of the thesis gets a definitive answer, depending on whether the Proponent can always win, that is, whether he has a winning strategy. We have introduced the basic notions related to this level in the previous section. See also the end of the next section, as well as Chaps. 3 and 5 for more explanations.

Now that the dialogical account of formation rules has been clarified, we may further develop our analysis of plays by introducing play-objects.

2.3 Play-Objects

The idea now is to design dialogical games in which the players' posits are of the form "$p : \varphi$" and get their meaning by the way they are used in the game: how they are challenged and defended. This requires analysing the form of a given play-object p, which depends on φ, and how a play-object can be obtained from other, simpler, play-objects. The standard dialogical semantics (Sect. 2.1) for logical constants gives us the information we need. The main logical constant of the expression at stake provides the basic information as to what a play-object for that expression consists of:

> A play for $\mathbf{X}\,!\,\varphi \vee \psi$ is obtained from two plays p_1 and p_2, in which p_1 is a play for $\mathbf{X}\,!\,\varphi$ and p_2 is a play for $\mathbf{X}\,!\,\psi$. According to the standard dialogical approach to disjunction, the player $\mathbf{X}$ is the one who can switch from p_1 to p_2 and conversely.
>
> A play for $\mathbf{X}\,!\,\varphi \wedge \psi$ is obtained similarly, except that the player $\mathbf{Y}$ is the one who can switch from p_1 to p_2.
>
> A play for $\mathbf{X}\,!\,\varphi \rightarrow \psi$ is obtained from two plays p_1 and p_2, in which p_1 is a play for $\mathbf{Y}\,!\,\varphi$ and p_2 is a play for $\mathbf{X}\,!\,\psi$. The player $\mathbf{X}$ is the one who can switch from p_1 to p_2.
>
> The standard dialogical particle rule for negation rests on the interpretation of $\neg\varphi$ as an abbreviation for $\varphi \rightarrow \bot$, although it is usually left implicit. From this follows that one obtains plays for $\mathbf{X}\,!\,\neg\varphi$ in a way similar to plays for a material implication, that is from two plays p_1 and p_2 in which p_1 is a play for $\mathbf{Y}\,!\,\varphi$, p_2 is a play for $\mathbf{X}\,!\,\bot$, and $\mathbf{X}$ can switch from p_1 to p_2. Notice that this approach covers the standard game-theoretical interpretation of negation as role-switch: p_1 is a play for a $\mathbf{Y}$-move.

As for quantifiers, a detailed discussion will be given after the particle rules. We would like to point out for now that, just like what is done in CTT, we are dealing with quantifiers for which the type of the bound variable is always specified. We thus consider expressions of the form $(Qx : A)\varphi$, where Q is a quantifier symbol.

Table 2.5, on p. 22, presents the particle rules.

Let us point out that we have added a challenge of the form $\mathbf{Y}\ ?_{prop}$ by which the challenger questions the fact that the expression at the right-hand side of the semi-colon is a proposition. This connects back with the formation rules of Sect. 2.2 via $\mathbf{X}$'s defence. Further details will be given in the discussion following the structural rules.

It may happen that the form of a play-object is not explicit at first. In such cases we deal with expressions of the form (for example) "$p : \varphi \wedge \psi$". In the relevant challenges and defences, we then use expressions such as $L^{\wedge}(p)$ and $R^{\wedge}(p)$ used in our example. We call these expressions *instructions*. Their respective interpretations are "take the left part of p" and "take the right part of p". In instructions we indicate the logical constant at stake.[14] This keeps the formulations explicit enough, in particular in the case of embedded instructions. We must keep in mind the important differences between play-objects depending on the logical constant. Consider for example the cases of conjunction and disjunction:

[14] If needed, we use subscripts to prevent scope ambiguities in the case of embedded occurrences of the same quantifier.

Table 2.5 Particle rules

Posit	Challenge	Defence
$\mathbf{X}\,!\,\varphi$ (where no play-object has been specified for φ)	$\mathbf{Y}\,?\,play\text{-}object$	$\mathbf{X}\,!\,p : \varphi$
$\mathbf{X}\,!\,p : \varphi \vee \psi$	$\mathbf{Y}\,?_{prop}$	$\mathbf{X}\,!\,\varphi \vee \psi : prop$
	$\mathbf{Y}\,?\,[\varphi/\psi]$	$\mathbf{X}\,!\,L^{\vee}(p) : \varphi$ or $\mathbf{X}\,!\,R^{\vee}(p) : \psi$ **[the defender has the choice]**
$\mathbf{X}\,!\,p : \varphi \wedge \psi$	$\mathbf{Y}\,?_{prop}$	$\mathbf{X}\,!\,\varphi \wedge \psi : prop$
	$\mathbf{Y}\,?\,L$ or $\mathbf{Y}\,?\,R$ **[the challenger has the choice]**	$\mathbf{X}\,!\,L^{\wedge}(p) : \varphi$ respectively $\mathbf{X}\,!\,R^{\wedge}(p) : \psi$
$\mathbf{X}\,!\,p : \varphi \rightarrow \psi$	$\mathbf{Y}\,?_{prop}$	$\mathbf{X}\,!\,\varphi \rightarrow \psi : prop$
	$\mathbf{Y}\,!\,L^{\rightarrow}(p) : \varphi$	$\mathbf{X}\,!\,R^{\rightarrow}(p) : \psi$
$\mathbf{X}\,!\,p : \neg\varphi$	$\mathbf{Y}\,?_{prop}$	$\mathbf{X}\,!\,\neg\varphi : prop$
	$\mathbf{Y}\,!\,L^{\perp}(p) : \varphi$	$\mathbf{X}\,!\,R^{\perp}(p) : \perp$
$\mathbf{X}\,!\,p : (\exists x : A)\varphi$	$\mathbf{Y}\,?_{prop}$	$\mathbf{X}\,!\,(\exists x : A)\varphi : prop$
	$\mathbf{Y}\,?\,L$ or $\mathbf{Y}\,?\,R$ **[the challenger has the choice]**	$\mathbf{X}\,!\,L^{\exists}(p) : A$ respectively $\mathbf{X}\,!\,R^{\exists}(p) : \varphi(L^{\exists}(p))$
$\mathbf{X}\,!\,p : \{x : A \mid \varphi\}$	$\mathbf{Y}\,?\,L$ or $\mathbf{Y}\,?\,R$ **[the challenger has the choice]**	$\mathbf{X}\,!\,L^{\{\dots\}}(p) : A$ respectively $\mathbf{X}\,!\,R^{\{\dots\}}(p) : \varphi(L^{\{\dots\}}(p))$
$\mathbf{X}\,!\,p : (\forall x : A)\varphi$	$\mathbf{Y}\,?_{prop}$	$\mathbf{X}\,!\,(\forall x : A)\varphi : prop$
	$\mathbf{Y}\,!\,L^{\forall}(p) : A$	$\mathbf{X}\,!\,R^{\forall}(p) : \varphi(L^{\forall}(p))$
$\mathbf{X}\,!\,p : B(k)$ (for atomic B)	$\mathbf{Y}\,?_{prop}$	$\mathbf{X}\,!\,B(k) : prop$
	$\mathbf{Y}\,p : B(k)?$	$\mathbf{X}\,sic(n)$ (**X indicates that Y posited it at move** n)

- A play-object p for a disjunction is composed of two play-objects, but each of them constitutes a sufficient play-object for the disjunction. Moreover it is the defender who makes the choice between $L^{\vee}(p)$ and $R^{\vee}(p)$.
- A play-object p for a conjunction is also composed of two play-objects, but this time the two of them are necessary to constitute the play-object for the conjunction.

> It is then the challenger's privilege to ask for either or both (provided the other rules allow him to do so).[15]

Accordingly, $L^{\wedge}(p)$ and $L^{\vee}(p)$, say, are actually different things and the notation takes that into account.

Let us now focus on the quantifier rules. Dialogical semantics highlights the fact that there are two distinct moments when considering the meaning of quantifiers: choosing a suitable substitution term for the bound variable, and instantiating the formula after replacing the bound variable with the chosen substitution term. However, the standard dialogical approach presupposes a unique, global, collection of objects over which the quantifiers range. Things are different with the explicit language borrowed from CTT. Quantification is always relative to a set, and there are sets of many different kinds of objects (sets of individuals, sets of pairs, sets of functions, etc.). Owing to the instructions we can give a general form for the particle rules, and the object is specified in a third and later moment when instructions are "resolved" by means of the structural rule $SR4.1$ presented in the next section.

Constructive Type Theory clearly shows the basic similarity there is, as soon as propositions are thought of as sets, between conjunction and existential quantifier on the one hand and material implication and universal quantifier on the other hand. Briefly, the point is that they are formed in similar ways and their elements are generated by the same kind of operations.[16] In our approach, this similarity manifests itself in the fact that a play-object for an existentially quantified expression is of the same form as a play-object for a conjunction. Similarly, a play-object for a universally quantified expression is of the same form as one for a material implication.[17]

The particle rule just before the one for universal quantification is a novelty in the dialogical approach. It involves expressions commonly used in Constructive Type Theory to deal with separated subsets. The idea is to understand those elements of A *such that* φ as expressing that at least one element $L^{\{\dots\}}(p)$ of A witnesses $\varphi(L^{\{\dots\}}(p))$. The same correspondence that linked conjunction and existential quantification now appears.[18] This is not surprising since such posits actually have an

---

[15] See in particular in the next section the repetition ranks in the structural rule $SR1i$.

[16] More precisely, conjunction and existential quantifiers are two particular cases of the Σ operator (disjoint union of sets), whereas material implication and universal quantifiers are two particular cases of the Π operator (indexed product on sets). See for example Ranta (1994, Chap. 2).

[17] Still, if we are playing with classical structural rules, there is a slight difference between material implication and universal quantification which we take from Ranta (1994, Table 2.3)., namely that in the second case p_2 always depends on p_1.

[18] As pointed out in Martin-Löf (1984), subset separation is another case of the Σ operator. See in particular p. 53:

> Let A be a set and $B(x)$ a proposition for $x \in A$. We want to define the set of all $a \in A$ such that $B(a)$ holds (which is usually written $\{x \in A : B(x)\}$). To have an element $a \in A$ such that $B(a)$ holds means to have an element $a \in A$ together with a proof of $B(a)$, namely an element $b \in B(a)$. So the elements of the set of all elements of A satisfying $B(x)$ are pairs (a, b) with $b \in B(a)$, i.e. elements of $(\Sigma x \in A)B(x)$. Then the Σ-rules play the role of the Comprehension Axiom (or the separation principle in ZF).

existential aspect: in $\{x : A \mid \varphi\}$ the left part "$x : A$" signals the existence of a play-object. Let us point out that since the expression stands for a set, it is not presupposed to be a proposition in **X**'s posit. This is why it cannot be challenged with the request "$?_{prop}$".

As we previously said, in the dialogical approach to CTT every object is known as instantiating a type and this constitutes the most elementary form of assertion $a : A$. Furthermore, instructions are in fact substitution commitments in a sense very close to the one mentioned by Brandom.[19] A thorough study is yet to be done on the substitutional approach to subsentential expressions and the role of instructions, though in our view it would be necessary for the exploration of both the formal consequences of Brandom's insights and the philosophical tenets underlying the notion of instruction.

Let us now consider the rule for the elementary case so as to close on the particle rules and complete our remarks of Sect. 2.1 on Rückert's point about legitimacy (*Geltung*). In this rule, but also in the associated formation rule of Sect. 2.2, the defence "$sic(n)$" recalls that the adversary has previously made the same posit. The rule works in a similar fashion as the Formal Rule of the standard formulation of Sect. 2.1, except that it is applicable to both players: it is not limited to the Proponent. We say similar in the sense that the rule allows players to perform a kind of copy-cat. Once that aspect of the Formal Rule is provided, we can work with a modified version of the rule which we will introduce with more explanations in the next section.

Despite the similarity we have just mentioned, there is a crucial difference with standard dialogical games. Elementary sentences are associated with play-objects, and one such sentence can be associated with many different play-objects in actual courses of the game. Therefore, and this is a most important point, the defence "$sic(n)$" does not express a copy-cat on the elementary sentence alone, but on the whole posit. We thus have a game rule such that, for a given elementary sentence, there are as many ways to give reasons for it (to defend it) as there are play-objects for it. Formulating the rule with the defence "$sic(n)$" is very different from merely integrating the standard Formal Rule at the local level: "$sic(n)$" is an abbreviation useful to provide an abstract rule, but because play-objects are introduced, it actually embodies a fully fledged semantics in terms of asking for and giving reasons.

So far, apart from the rule for subset-separation and the rule for elementary sentences, we have mostly adapted the rules of standard dialogical games to the explicit language we are working with. Now because of the explicit nature of this language,

---

[19]See Brandom (1994, pp. 425–426):

> So for an expression to be used as a singular term, there must be **some** substantive substitutional commitment undertaken by the one who uses it. It is not necessary that either the one who undertakes that commitment or the one who attributes it—by attributing a doxastic commitment that would be avowed by the assertion of a sentence containing the singular term—be able to specify just what the content of that commitment is. [...]
>
> Purported reference to objects must be understood in terms of substitutional commitments linking diverse expressions.

there are more rules related to the meaning explanations of play-objects and types. The next rules involve what is known in CTT as definitional equality. These rules introduce a different kind of provisional clause, namely a clause in which the defender is the player committed to the expression within the clause and thus he, rather than the challenger, will eventually posit it. In standard CTT there is no need for such a distinction since there are no players. However, in dialogical games the distinction can and must be made depending on who posits the proviso. Accordingly we use the notation $< \cdots >$ to signal that it is the player making the posit who is committed to the expression in the proviso clause and $(\cdots)$ when it is the challenger.

We have already considered the latter case in this chapter. As for the first case, let π be a posit and $< \cdots >$ a proviso which the utterer is committed to. The general form of the rule is then the following:

Posit	Challenge	Defence
$\mathbf{X}\,!\pi < \cdots >$	$\mathbf{Y}\,?_{[\pi]}$	$\mathbf{X}\,![\pi]$
	or	
	$\mathbf{Y}\,?_{[<>]}$	$\mathbf{X}\,![<>]$
	where $?_{[\pi]}$ and $![\pi]$ stand respectively for the relevant challenge or defence against π, and similarly for $?_{[<>]}$ and $![<>]$	

In the initial posit, **X** commits himself to both π and the proviso. Hence **Y** is entitled to question either one, and he is the one to choose which to ask for. The rule states that the challenger can question either part of the initial posit, and that in each case he does so depending on the form of the expression. An illustration is helpful here. Assume the initial posit is $p : (\forall x : A)B(x) < c : C >$ which reads "given $c : C$ we have $B(x)$ for all $x : A$; and the player making the posit commits himself to the proviso". Then the rule is applied in the following way:

Posit	Challenge	Defence
$\mathbf{X}\,!p : (\forall x : A)B(x) < c : C >$	$\mathbf{Y}\,!L^{\forall}(p) : A$	$\mathbf{X}\,!R^{\forall}(p) : B(L^{\forall}(p))$
	or	
	$\mathbf{Y}\,c : C?$	$\mathbf{X}\,sic(n)$

In this case, the challenger can attack either part of the initial posit. To challenge the first part, he applies the particle rule for universal quantification. The second part is challenged by applying the particle rule for elementary posits.

A typical case in which provisos of the form $< \cdots >$ occur is functional substitution. Assume some function f has been introduced, for example with $f(x) : B(x : A)$. When a player uses $f(a)$ in a posit, for some $a : A$, the antagonist is entitled to ask him what the output of f is, given a as input. Now $f(a)$ can be used either at the left or at the right of the colon. Accordingly we have two rules:

(Function-substitution)

Posit	Challenge	Defence
$\mathbf{X}\,!f(a):\varphi$	$\mathbf{Y}\ f(a)/?_{<=>}$	$\mathbf{X}\,!f(a)/k_i:\varphi < f(a)=k_i : B >$
$\mathbf{X}\,!\alpha:\varphi[f(a)]$	$\mathbf{Y}\ f(a)/?_{<=>}$	$\mathbf{X}\,!\alpha:\varphi[f(a)/k_i] < \varphi[f(a)]=\varphi[f(a)/k_i]: set >$

The subscript '<=>' in the challenges indicates that the substitution is related to some equality, and the defender endorses an equality in the proviso of the defence. The second rule—in which α can be a play-object or an instruction—is applied in the dialogical take on the Axiom of Choice. See the "second play" in Sect. 4.1.1.

Important remark: These two rules express a double commitment for the defender who is committed to the proviso in the defence. One might therefore argue that the rules could also be formulated as involving two challenges (and two defences). There are however two problems with such an approach. For illustration purposes, let us consider such a formulation of the second rule involving two steps:

Posit	Challenge	Defence
$\mathbf{X}\,!\alpha:\varphi[f(a)]$	$\mathbf{Y}\,L(f(a))/?$	$\mathbf{X}\,!\,p:\varphi[f(a)/k_i]$
	$\mathbf{Y}\,R(f(a))/?$	$\mathbf{X}\,!\varphi[f(a)]=\varphi[f(a)/k_i]:set$

The first problem is that the second challenge works as if the proviso $\varphi[f(a)] = \varphi[f(a)/k_i] : set$ was implicit in the initial posit and had to be made explicit. However this is a slightly misguided approach since the proviso does not concern the initial posit: the proviso must be established only after $\mathbf{X}$ has chosen k_i for the substitution. The second problem is related to the first: in such a formulation the challenger is the one who can choose between asking $\mathbf{X}$ to perform the substitution and asking him to posit the proviso. It thus allows the challenger to perform just the second challenge without asking for the substitution, which brings us back to the first problem. Moreover, introducing a choice for one of the players results, when the rule can be applied, in multiplying the number of alternative plays (in particular when the repetition rank of the challenger is 1). For all these reasons, such an alternative formulation is less satisfactory than the one we gave above.

Functional substitution is closely related to the Π-Equality rule, which we now introduce together with Σ.

(Π-Equality) We use the CTT notation Π which covers the cases of universal quantification and material implication.

Posit	Challenge	Defence
$\mathbf{X}\,!p:(\Pi x:A)\varphi$		
$\mathbf{Y}\,!L^{\Pi}(p)/a:A$		
$\mathbf{X}\,!R^{\Pi}(p):\varphi(a/x)$	$\mathbf{Y}\ ?_{\Pi\text{-}Eq}$	$\mathbf{X}\,!p(a)=R^{\Pi}(p):\varphi(a/x)$

(Σ-Equality) The rule is similar for existential quantification, subset separation, and conjunction. Thus we use the notation from CTT with the Σ operator. In the following rule I^{Σ} can be either L^{Σ} or R^{Σ}, and i can be either 1 or 2: it is 1 when I is L and 2 when I is R.

Posit	Challenge	Defence
$\mathbf{X}\,!p : (\Sigma x : \varphi_1)\varphi_2$ $\mathbf{Y}\, I^{\Sigma}(p)/?$ $\mathbf{X}\,!p_i/I^{\Sigma}(p) : \varphi_i$	$\mathbf{Y}\,?_{\Sigma\text{-}Eq}$	$\mathbf{X}\,!I^{\Sigma}(p) = p_i : \varphi_i$

Notice that these rules have several preconditions: there is no lone initial posit triggering the application of the rule. From a dialogical perspective, these rules intend to allow the challenger to take advantage of information from the history of the current play—including resolutions of instructions—to make **X** posit some equality. For an application, see the second play in Sect. 4.1.1 in which the Π-Equality rules play a prominent role.

These rules strongly suggest a close connection between the CTT equality rules for logical constants and the dialogical instructions through what we will call in the next section their *resolution*. It is thus important to remind the significant differences between them, and especially that the particle rules define operations on propositions that are very different from the set-theoretical operations in CTT.

Let us discuss this topic before giving the remaining rules. The main point involves the *player-independence* of the rules, which we have only mentioned at the beginning of this chapter. By this we refer to the fact that the rules presented in this section are the same for the two players, which is why they are formulated with the variables **X** and **Y**. Various publications[20] have already linked the notion of player-independence to the immunity of the dialogical framework to different trivializing connectives such as Prior's *tonk*[21] and thus more generally to Dummett's requirement of *harmony* between Introduction and Elimination rules.[22]

In CTT, the harmonious relationship between Introduction and Elimination rules is made explicit by associating each logical constant with a suitable Equality rule (Sect. 1.2.2).[23] More precisely, the possibility to have such equality rules ensures harmony between Introduction and Elimination rules.

But at the same time the triplets Introduction-Elimination-Equality rules in the usual presentation of the CTT framework reinforces a certain form of asymmetry between Introduction and Elimination rules. The very idea of requiring rules to be harmonious advocates for the possibility of an approach with rules of only one kind (either Introduction or Elimination), and to design harmonious corresponding rules of the other kind—see Dummett (1993). By giving priority to Introduction rules, Gentzen (1934–1935) already observed one direction in this possible alternative,

[20] See Rahman and Keiff (2005), Rahman et al. (2009), Rahman (2012).

[21] Cf. Prior (1961).

[22] See Dummett (1973).

[23] They are, after all, linearized versions of Prawitz's reduction steps in Natural Deduction. See Sundholm (1997, Sect. 3.6).

and in this respect the standard presentation of CTT follows him: Equality rules establish the harmony of Elimination rules with respect to previously given Introduction rules. That is to say, we can start with the Introduction rules, then derive from them corresponding Elimination rules and check that they are harmonious.[24]

Therefore the Equality rules in the standard presentation cannot be used to achieve the other alternative, namely to start with Elimination rules then look for possible corresponding Introduction rules. To our knowledge this program has not been addressed in full details, but some hints and suggestions have been made: a promising candidate to replace the standard Equality rules for this purpose is the η-conversion (see in particular Primiero (2008) on this). But then again, the result would be the converse of the standard presentation, still featuring one kind of rule as conceptually prior to the other.

In our view, the dialogical approach is a promising one because the rules, being player-independent, do not dichotomise Introduction and Elimination. A key point is that this also holds in the case of the Π and Σ-Equality rules which we have given above: the dialogical rules do not have this "one-sided" aspect of the CTT Equality rules or their candidate alternative, the η-conversion. Still, as we will see in Chaps. 3 and 5, the connection between the dialogical and the CTT approaches becomes salient when we consider *applications* (by the players) *of the dialogical rules at the level of strategies*. The **P**-application versus **O**-application thus not only gives us Introduction versus Elimination rules in the sense of CTT, but also two versions of the Π and Σ-Equality. Moreover, it seems like these two versions yield on the one hand the standard Equality rules of CTT, and on the other the alternative η-conversion: this looks promising in respect to Dummett's 1993 remark mentioned above. It is also the topic of ongoing studies on the notion of harmony using the dialogical approach to CTT.

Applying Π or Σ-Equality rules is particularly useful for the Proponent when he is the challenger (**Y**): they provide a way for the challenger to make **X** perform a particular posit. The rule is obviously interesting for **P** when it comes to elementary posits: by compelling **O** to make an elementary posit, **P** ensures he can defend his own version of the elementary posit with "$sic(n)$", should he need to do so. In other words, the Proponent can use Π and Σ-Equality to be able to resort to copy-cat so as to defend his elementary posits.

This suggests a close connection between the Equality rules of CTT and copy-cat in dialogical games. In fact, we could probably replace the two rules above with rules conveying the idea that when the three preconditions are met within a play, one of the players is entitled to make the posits appearing as defences in the two rules above. That is to say, he is allowed to make the posits $!p(a) = R^{\Pi}(p) : \varphi(a/x)$ and $!I^{\Sigma}(p) = p_i : \varphi_i$, respectively. In the form of a local rule with attack and defence. This yields,

---

[24]This is the path taken for example in Thompson (1991, Sect. 8.4).

Posit	Challenge	Defence
$\mathbf{X}\,!p(a) = R^{\Pi}(p) : \varphi(a/x)$	$\mathbf{Y}\,?$	$\mathbf{X}\,\Pi - Eq - sic(j, k, l)$ where j, k, l are the moves at which the three preconditions have been played

Likewise for Σ-Eq. Such a formulation insists on the link with copy-cat by using the abbreviation "$sic(j, k, l)$", comparable to "$sic(n)$" in the case of elementary posits.

Formulating the rule in such a way comes quite naturally to the game perspective in which copy-cat is a well-established notion. It nevertheless gets more difficult when going in depth. We have left the challenge underspecified in this rule: it may not be obvious that the initial posit by **X** is related to a previous Π-expression (i.e. involves a universal or a material implication). In fact it is precisely the point of the defence to make it salient. But then what is there in the posit entitling a challenge that can be answered with the abbreviation "sic"? And what would such a challenge look like? To sum up, it is not easy to produce a convincing and explicit rule.

The first versions we gave, the ones in which three preconditions entitle **Y** to ask **X** to perform a certain posit in accordance to Π and Σ-Equality, look more promising because it can be of great importance to have rules as explicit as possible. For example, it is closely related to the way extension and intension are distinguished in CTT—see an application with the discussion on the differences between the intensional and extensional versions of the Axiom of Choice in Chap. 4.

These few remarks are surely not enough to deal with the different but related topics we have just mentioned. If anything, they suggest that there are various directions in which the link between CTT and dialogical games should be further explored, well beyond the single technical result at stake in this study.

Let us stop the digression here and come back to the other rules involving equality.

(Reflexivity within set)

Posit	Challenge	Defence
$\mathbf{X}\,!\,A : set$	$\mathbf{Y}\,?_{set\text{-refl}}$	$\mathbf{X}\,!\,A = A : set$

(Symmetry within set)

Posit	Challenge	Defence
$\mathbf{X}\,!\,A = B : set$	$\mathbf{Y}\,?_{set\text{-symm}}$	$\mathbf{X}\,!\,B = A : set$

(Transitivity within set)

Posit	Challenge	Defence
$\mathbf{X}\,!\,A = B : set$ ⋮ $\mathbf{X}\,!\,B = C : set$	$\mathbf{Y}\,?_{set\text{-trans}}$	$\mathbf{X}\,!\,A = C : set$

(Reflexivity within A)

Posit	Challenge	Defence
$\mathbf{X} ! a : A$	$\mathbf{Y} ?_{A\text{-refl}}$	$\mathbf{X} ! a = a : A$

(Symmetry within A)

Posit	Challenge	Defence
$\mathbf{X} ! a = b : A$	$\mathbf{Y} ?_{A\text{-symm}}$	$\mathbf{X} ! b = a : A$

(Transitivity within A)

Posit	Challenge	Defence
$\mathbf{X} ! a = b : A$ ⋮ $\mathbf{X} ! b = c : A$	$\mathbf{Y} ?_{A\text{-trans}}$	$\mathbf{X} ! a = c : A$

(Set-equality/Extensionality)

Posit	Challenge	Defence
$\mathbf{X} ! A = B : set$	$\mathbf{Y} ! a : A$ $\mathbf{Y} ! a = b : A$	$\mathbf{X} ! a : B$ $\mathbf{X} ! a = b : B$

(Set-substitution)

Posit	Challenge	Defence
$\mathbf{X} ! B(x) : set(x : A)$	$\mathbf{Y} ! x = a : A$	$\mathbf{X} ! B(x/a) : set$
$\mathbf{X} ! B(x) : set(x : A)$	$\mathbf{Y} ! a = c : A$	$\mathbf{X} ! B(a) = B(c) : set$
$\mathbf{X} ! b(x) : B(x)(x : A)$	$\mathbf{Y} ! a : A$	$\mathbf{X} ! b(a) : B(a)$
$\mathbf{X} ! b(x) : B(x)(x : A)$	$\mathbf{Y} ! a = c : A$	$\mathbf{X} ! b(a) = b(c) : B(a)$

In these last rules, we have considered the simpler case in which there is only one assumption in the proviso or context. The rules can obviously be generalized for provisos featuring multiple assumptions.

This ends the presentation of the dialogical notion of play-object and of the rules which give an abstract description of the local proceedings of dialogical games. Next we consider the global conditions taking part in the development of dialogical plays.

2.4 The Development of a Play

We will deal in this section with the other kind of dialogical rules called structural rules. These rules govern the way plays globally proceed and are therefore an important aspect of dialogical semantics. We will work with the following structural rules:

SR0 (Starting Rule). Any dialogue starts with the Opponent positing initial concessions, if any, and the Proponent positing the thesis. After that the players each choose a positive integer called repetition rank.

SR1i (Intuitionisitic Development Rule). Players move alternately. After the repetition ranks have been chosen, each move is a challenge or a defence in reaction to a previous move, in accordance with the particle rules. The repetition rank of a player bounds the number of challenges he can play in reaction to a same move. Players can answer only against the *last non-answered* challenge by the adversary.[25]

SR2 ("Priority to Formation" Rule). **O** starts by challenging the thesis with the request '$?_{prop}$'. The game then proceeds by applying the formation rules first so as to check that the thesis is indeed a proposition. After that the Opponent is free to use the other local rules insofar as the other structural rules allow it.

SR3 (Modified Formal Rule). **O**'s elementary sentences cannot be challenged. However, **O** can challenge a **P** elementary move provided she did not herself play it before.

Since we have particle rules for elementary sentences involving the defence "$sic(n)$" we have no need for a Formal Rule which entitles a player to copy-cat some moves of the adversary.[26] We must however also ensure that the strictly internal aspect related to the idea of *Geltung* in the dialogical approach to meaning is not lost, and that the asymmetry between the player **P** who brings forward the thesis and his adversary **O** is accounted for. This is why the standard Formal Rule is replaced by this modified version.

SR4.1 (Resolution of Instructions). Whenever a player posits a move in which instructions $I_1, \ldots, I_n$ occur, the other player can ask him to replace these instructions (or some of them) by suitable play-objects.

If the instruction (or list of instructions) occurs at the right of the colon and the posit is the tail of an universally quantified sentence or of an implication (so that these instructions occur at the left of the colon in the posit of the head of the implication), then it is the challenger who can choose the play-object. In these cases the player who challenges the instruction is also the challenger of the universal quantifier and/or of the implication.

---

[25] This last clause is known as the *Last Duty First* condition, and is the clause making dialogical games suitable for Intuitionistic Logic, hence the name of this rule.

[26] But let us insist once more on the important point we raised in Sect. 2.3: contrary to standard dialogical games, copy-cat does not apply only to elementary sentences but also to posits in which such sentences are associated with play-objects.

Otherwise it is the defender of the instructions who chooses the suitable play-object:

Posit	Challenge	Defence
$\mathbf{X}\,\pi(I_1, \ldots, I_n)$	$\mathbf{Y}\, I_1, \ldots, I_m/?\,(m \leq n)$	$\mathbf{X}\,\pi(b_1, \ldots, b_m)$ If the instruction occurring at the right of the colon is the tail of either a universal or an implication (such that $I_1, \ldots, I_n$ also occurs at the left of the colon in the posit of the head), then $b_1, \ldots, b_m$ was previously chosen by the challenger Otherwise the defender chooses

Important remark. In the case of embedded instructions $I_1(\ldots(I_k)\ldots)$, the substitutions are thought of as being carried out from I_k to I_1: first substitute I_k with some play-object b_k, then $I_{k-1}(b_k)$ with b_{k-1} etc. until $I_1(b_2)$. If such a progressive substitution has already been carried out once, a player can then replace $I_1(\ldots(I_k)\ldots)$ directly.

SR4.2 (Substitution of Instructions). When, during the play, the play-object b has been chosen by any of the two players for an instruction I, and player **X** makes any posit $\pi(I)$, then the other player can ask for I to be substituted by b in this posit:

Posit	Challenge	Defence
$\mathbf{X}\,\pi(I)$ (where I/b has been previously established)	$\mathbf{Y}\,?\,I/b$	$\mathbf{X}\,\pi(b)$

The idea is that the resolution of an instruction yields a certain play-object for some substitution term, and therefore the same play-object can be assumed to result from any other occurrence of the same substitution term: instructions, after all, are functions and must yield as such the same play-object for the same substitution term.

SR5 (Winning Rule for plays). For any p, a player who posits "$p : \bot$" looses the current play. Otherwise the player who makes the last move in a dialogue wins it.

In comparison to the rules of standard dialogical games, some additions in the rules we just gave have been made, namely $SR2$ and $SR4.1$-2. Also, the so-called Formal Rule (here $SR3$) and the Winning Rule are a bit different. Since we made explicit the use of $\bot$ in our games, we need to add a rule for it: the point is that positing falsum leads to immediate loss. We could say that it amounts to a withdrawal.[27] Hence the formulation of the Winning Rule for plays above.

We need the rules $SR4.1$ and $SR4.2$ because of some features of CTT's explicit language. In CTT it is possible to account for questions of dependency, scope, etc.

[27] See Keiff (2007).

directly at the language level. In this way various puzzles, such as anaphora, get a convincing and successful treatment. The typical example, considered below, is the so-called donkey sentence "Every man who owns a donkey beats it". The two rules account for the way play-objects can be ascribed to what we have called instructions. See the dialogue in Sect. 2.5 for an application.

The rule $SR2$ is consistent with the common practice in CTT to start demonstrations by checking or establishing the formation of propositions before proving their truth. Notice that this step also covers the formation of sets—membership, generation of elements, etc.—occurring in hypothetical posits and in quantifiers. In the current study, however, we can overlook this rule: we can take it for granted that expressions are well formed because we have restricted this study to the valid fragment of CTT. That is to say, we will only consider cases for which it is not necessary to carry out the formation steps since even if they were carried out, the players would always be able to justify that their expressions are well formed. We will, for this, always take examples in which good formation is guaranteed by hypotheses introduced as initial concessions by the Opponent at the beginning of the play.

What is more, it seems like we could liberalise the rule $SR2$. But because of the number of rules we have introduced, verifying this carefully is a delicate task that we will not carry out in this study. Let us for now simply mention that it seems sensible enough in dialogues to combine more freely the process linked to formation rules with the other rules at stake in the development of a play. Questioning the status of expressions as they are introduced in the course of the game does in fact seem perfectly consistent with actual practices. Suppose for example that player **P** has posited '$p : \varphi \vee \psi$'. As soon as he has posited that the disjunction is a proposition—i.e. as soon as he has posited '$\varphi \vee \psi : prop$'—the other player knows how to challenge the disjunction and should be free to either keep on exploring the formation of the expression or to challenge the first posit. The point is that in a way it makes more sense to check whether φ is a proposition or not once (or if) **X** posits it in order to defend the disjunction. Doing so in a 'monological' framework such as CTT would probably bring various confusions, but the dialogical approach to meaning should quite naturally allow this additional dynamic aspect. Nonetheless, in order to generalise the equivalence result beyond the valid fragment of CTT (the reason why we have introduced rule $SR2$), it seems sensible in our view to clearly distinguish in a fashion close to CTT the steps linked to the formation from the other aspects of meaning.

The definitions of plays, games and strategies are the same as those given in Sect. 2.1. Let us now recall them. A play for φ is a sequence of moves in which φ is the thesis posited by the Proponent and which complies with the game rules. The dialogical game for φ is the set of all possible plays for φ and its extensive form is nothing but its tree representation. Thus, every path in this tree which starts with the root is the linear representation of a play in the dialogical game at stake.

We say that a play for φ is terminal when the last move is made by player **X** and there is no further move for **Y**. A strategy for player **X** in a given dialogical game is a function which assigns a legal **X**-move to each non-terminal play where it is **X**'s turn to move. When the strategy is a winning one for **X**, the assignation results in terminal plays won by **X**. It is common practice to consider in an equivalent way an **X**-strategy s as the set of terminal plays resulting when **X** plays according to s. The extensive form of s is then the tree representation of that set—which is by the way a fragment of the extensive form of the dialogical game. For more explanations on these notions, see Clerbout (2014c). The equivalence result between dialogical games and CTT is established by procedures of translation between a certain part of an extensive form of winning **P**-strategies—and more precisely what we call their core—and CTT demonstrations. We will give more details on how to isolate cores of strategies in Chap. 3.

2.5 Example

We end this presentation of dialogical games with an illustration of the approach. The example comes from Rahman et al. (2014) and consists in a dialogue in which the famous donkey sentence "Every man who owns a donkey beats it" is involved.

In his 1986 paper, G. Sundholm thoroughly discussed this famous puzzle in the context of Constructive Type Theory. As is well-known, the problem is to give a way to capture the back-reference of the pronoun "it". The point in Sundholm (1986) is that the explicit language of CTT makes it possible to express and account for such dependencies as soon as one pays attention to the fact that "a man who owns a donkey" is a member of the set

$$\{x : M|(\exists y : D)O(x, y)\}$$

For the detailed explanation of the CTT approach, see Sundholm's paper. The point we are interested in here is that the back-reference of the pronoun is dealt with in a similar way by using dialogical instructions. That is, we write the donkey sentence as

$$\big(\forall z : \{x : M|(\exists y : D)O(x, y)\}\big)B(L^{\{\ldots\}}(z), L^{\exists}(R^{\{\ldots\}}(z)))$$

where M is the set of men, D is the set of donkeys, $O(x, y)$ stands for "x owns y" and $B(x, y)$ stands for "x beats y".

Table 2.6 presents a dialogue in which this sentence occurs as one of the initial concessions by the Opponent.

Table 2.6 A dialogue involving the donkey sentence

	O			P	
I	$M : set$				
II	$D : set$				
III	$O(x,y) : set(x : M, y : D)$				
IV	$B(x,y) : set(x : M, y : D)$				
V	$!p : (\forall z : \{x : M \mid (\exists y : D)O(x,y)\})B(L^{\{\cdots\}}(z), L^{\exists}(R^{\{\cdots\}}(z)))$				
VI	$!m : M$				
VII	$!d : D$				
VIII	$!p' : Omd$				
				$!B(m,d)$	0
1	n := ...			m := ...	2
3	$?play\text{-}object$	(0)		$!q : B(m,d)$	30
25	$!R^{\forall}(p) : B(L^{\{\cdots\}}(z), L^{\exists}(R^{\{\cdots\}}(z)))$		(*V*)	$!L^{\forall}(p) : \{x : M \mid (\exists y : D)O(x,y)\}$	4
5	$L^{\forall}(p)/?$	(4)		$!z : \{x : M \mid (\exists y : D)O(x,y)\}$	6
7	$?_L$	(6)		$!L^{\{\cdots\}}(z) : M$	8
9	$L^{\{\cdots\}}(z)/?$	(8)		$!m : M$	10
11	$?_R$	(6)		$!R^{\{\cdots\}}(z) : (\exists y : D)Omy$	12
13	$R^{\{\cdots\}}(z)/?$	(12)		$(L^{\exists}(R^{\{\cdots\}}(z)), R^{\exists}(R^{\{\cdots\}}(z))) : (\exists y : D)Omy$	14
15	$L^{\exists}(R^{\{\cdots\}}(z))/?, R^{\exists}(R^{\{\cdots\}}(z))/?$	(14)		$!(d,p') : (\exists y : D)Omy$	16
17	$?_L$	(16)		$!L^{\exists}(d,p') : D$	18
19	$L^{\exists}(R^{\{\cdots\}}(z))/?$	(18)		$!d : D$	20
21	$?_R$	(16)		$!R^{\exists}(d,p') : Omd$	22
23	$R^{\exists}(R^{\{\cdots\}}(z))/?$	(22)		$!p' : Omd$	24
27	$!R^{\forall}(p) : B(m,d)$		(25)	$L^{\{\cdots\}}(z)/m, L^{\exists}(R^{\{\cdots\}}(z))/d$	26
29	$!q : B(m,d)$		(27)	$R^{\forall}(p)/?$	28

Explanations:

- Moves I–VIII. These moves are **O**'s initial concessions. Moves I to IV deal with the formation of expressions. After that the Opponent concedes the donkey sentence and atomic expressions related to the sets M, D and $O(x, y)$.
- Moves 0-3. The Proponent posits the thesis. The players choose their repetition ranks in moves 1 and 2. The actual value they choose does not really matter for the point we illustrate here: we simply assume they are sufficient for this play and leave them unspecified. When **P** posited the thesis he did not specify a play-object so **O** asks for it in move 3.
- Move 4. The Proponent chooses to launch a counter-attack by challenging the donkey sentence which **O** conceded at V. The rules do allow him to directly answer to the challenge, but then he would not be able to win.[28]
- Moves 5-24. The dialogue then proceeds in a rather straightforward way by applications of the rules introduced in Sects. 2.3 and 2.4. More precisely, this dialogue displays the case where **O** chooses to challenge **P**'s posits as much as she can before answering to **P**'s challenge 4.
  Notice that the Opponent cannot challenge the Proponent's atomic expressions posited at moves 10, 20 and 24. Since **O** made the same posits in her initial concessions VI to VIII, the Modified Formal Rule $SR3$ forbids her to challenge them.
- Move 25. When there is nothing left for her to challenge, **O** comes back to the last unanswered challenge by **P** which was move 4 and makes the relevant defence according to the particle rule for universal quantification.
- Moves 26-27. The resolution for instructions $L^{\{\cdots\}}(z)$ and $L^{\exists}(R^{\{\cdots\}}(z))$ have been carried out during the dialogue with moves 9-10 and moves 23-24. The Proponent can thus use

[28]Indeed, after he answers challenge 3 he has to defend the atomic posit played in defence. To successfully do so, he must make **O** perform the same posit. But the reader can check that in this case nothing would compel **O** to choose the same play-object. Hence, the way to victory for **P** is to let **O** choose a play-object for $B(m, d)$ first as in this dialogue.

the established substitutions to challenge move 25 according to the structural rule *SR*4.2. The Opponent defends by performing the requested substitutions.
- Moves 28-30. The Proponent then asks the play-object for which the instruction $R^{\forall}(z)$ stands. When she answers, the Opponent posits exactly what **P** needs to defend against **O**'s challenge 3. Notice that at this point this is the last unanswered challenge by **O**, therefore **P** is allowed to answer it in accordance to the structural rule *SR*1*i*. He does so with his move 30.
  Since **O** made the same posit, the rule *SR*3 forbids her to challenge it. She then has no further possible move, and the Proponent wins this dialogue.

The example illustrates the applications of many rules of the dialogical approach. Let us insist on the way the device of dialogical instructions enriches the language in order to account for dependency relations, such as anaphora, through their resolution. This is the reason why we chose an example involving the donkey sentence.

References

Brandom, R. 1994. *Making it Explicit*. Cambridge: Harvard UP.

Clerbout, N. 2014a. First-order dialogical games and tableaux. *Journal of Philosophical Logic* 43(4): 785–801. [Published first online 2013. doi:10.1007/s10992-013-9289-z].

Clerbout, N. 2014b. Finiteness of plays and the dialogical problem of decidability. *IfCoLog Journal of Logics and Their Applications* 1(1): 115–130.

Clerbout, N. 2014c. *La Sémantique Dialogique. Notions Fondamentales et Éléments de Metathéorie*. London: College Publications.

Clerbout, N., M.H. Gorisse, and S. Rahman. 2011. Context-sensitivity in Jain philosophy: A dialogical study of Siddharsigani's commentary on the handbook of logic. *Journal of Philosophical Logic* 40(5): 633–662.

Dummett, M. 1973. *Frege: Philosophy of Language*. London: Duckworth.

Dummett, M. 1993. Language and truth. In *The Seas of Language*, 117–165. Oxford: Clarendon Press.

Felscher, W. 1985. Dialogues as a foundation for intuitionistic logic. In *Handbook of Philosophical Logic*, vol. 3, ed. D. Gabbay and F. Guenthner, 341–372. Dordrecht: Kluwer.

Fiutek, V., H. Rückert, and S. Rahman. 2010. A dialogical semantics for Bonanno's system of belief revision. In *Constructions*, ed. P. Bour et al., 315–334. London: College Publications.

Fontaine, M. 2013. *Argumentation et Engagement Ontologique. Être, c'est être choisi*. London: College Publications.

Gentzen, G. 1934–1935. Untersuchungen über das logische Schliessen. *Mathematische Zeitschrift* 39: 176–210, 405–431. Also in M. Szabo, ed. tr., 1969, *The Collected Papers of Gehard Gentzen*. Amsterdam: North-Holland.

Kamlah, W., and P. Lorenzen. 1972. *Logische Propädeutik*, 2nd ed. Stuttgart and Weimar: Metzler.

Kamlah, W., and P. Lorenzen. 1984. *Logical Propaedeutic*. Lanham Md.: University Press of America. English translation of Kamlah and Lorenzen [1972] by H. Robinson.

Keiff, L. 2004a. Heuristique formelle et logiques modales non-normales. *Philosophia Scientiae* 8(2): 39–57.

Keiff, L. 2004b. Introduction à la dialogique modale et hybride. *Philosophia Scientiae* 8(2): 89–102.

Keiff, L. 2007. *Le Pluralisme Dialogique: Approches Dynamiques de l'Argumentation Formelle*. PhD thesis, Lille 3.

Keiff, L. 2009. Dialogical logic. *Stanford Encyclopedia of Philosophy*. http://plato.stanford.edu/entries/logic-dialogical/. Accessed 2013.

Lorenz, K. 2001. Basic objectives of dialogue logic in historical perspective. In *New Perspectives in Dialogical Logic*, ed. S. Rahman, and H. Rückert, 255–263. Springer (*Synthese* 127(1–2): special volume).

Lorenz, K. 2008. *Dialogischer Konstruktivismus*. Berlin: De Gruyter.

Lorenz, K. 2010a. *Logic, Language and Method: On Polarities in Human Experience*. Berlin: De Gruyter.

Lorenz, K. 2010b. *Philosophische Variationen: Gesammelte Aufsätze unter Einschluss gemeinsam mit Jürgen Mittelstraß geschriebener Arbeiten zu Platon und Leibniz*. Berlin: De Gruyter.

Lorenzen, P., and K. Lorenz. 1978. *Dialogische Logik*. Darmstadt: Wissenschaftliche Buchgesellschaft.

Lorenzen, P., and O. Schwemmer. 1975. *Konstruktive Logik, Ethik und Wissenschaftstheorie*, 2nd edn. Mannheim: Bibliographisches Institut.

Magnier, S. 2013. *Approche Dialogique de la Dynamique Épistémique et de la Condition Juridique*. London: College Publications.

Martin-Löf, P. 1984. *Intuitionistic Type Theory*. Naples: Bibliopolis. Notes by Giovanni Sambin of a series of lectures given in Padua, June 1980.

Popek, A. 2012. Logical dialogues from middle ages. In *Logic of Knowledge. Theory and Applications*, ed. C. Barés, 223–244. London: College Publications.

Primiero, G. 2008. *Information and Knowledge*. Dordrecht: Springer.

Prior, A. 1961. The runabout inference ticket. *Analysis* 21(2): 38–39.

Rahman, S. 1993. *Über Dialoge, Protologische Kategorien und andere Seltenheiten*. Frankfurt: P. Lang.

Rahman, S. 2009. A non normal logic for a wonderful world and more. In *The Age of Alternative Logics: Assessing Philosophy of Logic and Mathematics Today*, 2nd edn., ed. J. van Benthem et al., 311–334. Dordrecht: Springer.

Rahman, S. 2012. Negation in the logic of first degree entailment and tonk: A dialogical study. In *The Realism-Antirealism Debate in the Age of Alternative Logics*, ed. S. Rahman, G. Primiero, and M. Marion, 213–250. Dordrecht: Springer.

Rahman, S., and N. Clerbout. 2014. Constructive type theory and the dialogical turn. A new start for Erlangen constructivism. In *Dialogische Logik*, ed. J. Mittlestrass, and C. von Bülow, 127–184. Münster: Mentis.

Rahman, S., and L. Keiff. 2005. On how to be a dialogician. In *Logic, Thought, and Action*, ed. D. Vanderveken, 359–408. Dordrecht: Kluwer.

Rahman, S., and L. Keiff. 2010. La Dialectique entre logique et rhétorique. *Revue de Métaphysique et de Morale* 66(2): 149–178.

Rahman, S., and T. Tulenheimo. 2009. From games to dialogues and back: Towards a general frame for validity. In *Games: Unifying Logic, Language and Philosophy*, ed. O. Majer, 153–208. Dordrecht: Springer.

Rahman, S., N. Clerbout, and L. Keiff. 2009. On dialogues and natural deduction. In *Acts of Knowledge: History, Philosophy and Logic: Essays Dedicated to Göran Sundholm*, ed. G. Primiero, and S. Rahman, 301–336. London: College Publications.

Rahman, S., N. Clerbout, and Z. McConaughey. 2014. On play-objects in dialogical games. Towards a dialogical approach to constructive type theory. In *Modestly Radical or Radically Modest. Festschrift for Jean-Paul van Bendegem on the Occasion of his 60th Birthday*, ed. P. Allo, and B. van Kerkhove, 127–154. College Publications.

Ranta, A. 1994. *Type-Theoretical Grammar*. Oxford: Clarendon Press.

Redmond, J. 2010. *Logique Dynamique de la Fiction: Pour une Approche Dialogique*. London: College Publications.

Redmond, J., and M. Fontaine. 2011. *How to Play Dialogues: An Introduction to Dialogical Logic*. London: College Publications.

Rückert, H. 2011a. *Dialogues as a Dynamic Framework for Logic*. London: College Publications.

Rückert, H. 2011b. The conception of validity in dialogical logic. Talk at the workshop *Proofs and Dialogues*, Tübingen.

Sundholm, G. 1986. Proof-theory and meaning. In *Handbook of Philosophical Logic*, vol. 3, ed. D. Gabbay, and F. Guenthner, 471–506. Dordrecht: Reidel.
Sundholm, G. 1997. Implicit epistemic aspects of constructive logic. *Journal of Logic, Language and Information* 6(2): 191–212.
Thompson, S. 1991. *Type Theory and Functional Programming*. Addison-Wesley.
van Ditmarsch, H., W. van der Hoek, and B. Kooi. 2007. *Dynamic Epistemic Logic*. Berlin: Springer.

Chapter 3
From Dialogical Strategies to CTT Demonstrations

We now move to the demonstration of the left-to-right direction of the equivalence result. Let us assume that there is a winning **P**-strategy in the dialogical game for φ. We will take the extensive form (Sect. 2.1) of this strategy and present a procedure to extract from it what has been called in Rahman et al. (2009) its *core*. We will show after that how such a core can be transformed into a CTT demonstration of φ.

3.1 Towards the Core

The first step towards our goal is to ignore almost every possible choice of repetition rank for the Opponent. This can be done safely. Indeed,

> Assume there is a winning **P**-strategy in the dialogical game for φ. Let $\mathscr{D}_1(\varphi)$ denote the sub-game where the Opponent chooses her repetition rank to be $\mathtt{1}$. Then there is a winning **P**-strategy $s^\star$ in $\mathscr{D}_1(\varphi)$.[1]

Let us call $\mathbf{S}^\star$ the extensive form of $s^\star$.

3.1.1 Getting Rid of Infinite Ramifications

When she has to choose a play-object for a previously unresolved instruction, the Opponent often has infinitely many possible choices, though once she has chosen one

[1] See Clerbout (2014a, b): if there is a move by which the Opponent can force her victory, then nothing prevents her from playing it as soon as she has a chance to. Whether this move is a challenge or a defence, the repetition rank $\mathtt{1}$ is enough to allow her to play it in accordance with $SR1i$.

N. Clerbout and S. Rahman, *Linking Game-Theoretical Approaches with Constructive Type Theory*, SpringerBriefs in Philosophy,
DOI 10.1007/978-3-319-19063-1_3

she must keep it for the rest of the play (rule $SR4.2$). In such cases the play-object associated with the instruction will be a member of some set. Unless otherwise specified this set may be infinite. The Opponent can then choose among an infinite number of members when asked to replace the instruction with a play-object. Thus $\mathbf{S}^\star$ has infinitely many branches and we now rectify this.

Call a node t in $\mathbf{S}^\star$ *critical* if it has infinitely many *immediate* successors, and $S(t)$ the set of these successors. Our first aim is to partition $S(t)$, for each critical node in $\mathbf{S}^\star$, depending on the kind of moves associated with its members. Fortunately, since we are working with intuitionistic (i.e. with rule $SR1i$) dialogical games with play-objects having play-objects and instructions, the task will actually be simpler than with classical dialogues.[2] Let us describe the situation progressively:

1. Let us first recall that since we started with a **P**-strategy, branches are triggered by **O**-choices. Hence for every critical node t in $\mathbf{S}^\star$, the members of $S(t)$ are associated with **O**-moves. Each of these moves is obviously either a defence or a challenge.
2. Now, because the finiteness of plays is ensured by repetition ranks, the only way for t to be critical is for **O** to react in an infinite number of ways to at least one of its predecessors, or even t itself.
3. From the dialogical rules, we know that there are only two cases in which the Opponent has the local choice between infinitely many moves, namely:
   - when applying the Posit-Substitution rule of Sect. 2.2 to challenge a hypothetical move by instantiating the assumptions in the context. Indeed in this rule the challenger is the one choosing the instantiations of the x_is which occur in the assumptions; or
   - when applying the rule for Resolution of Instructions (rule $SR4.1$) as the defender, i.e. when choosing the play-objects to be substituted for given instructions.

4. Now, let us focus on nodes in $S(t)$ labelled with **O**-defences. Because we are dealing with intuitionistic dialogues we know that they are all answers to a unique **P**-challenge. Indeed the rule $SR1i$ states, among other things, the *last duty first* condition which we recall here:

   Players can answer only against the **last non-answered** challenge by the adversary.

   Call $Sd(t)$ the set of nodes in $S(t)$ associated with an **O**-defence. It follows from the previous remarks that if $Sd(t)$ is infinite then all of its members *are associated with the same moves modulo the play-object*.
5. Things are a bit less simple in the case of challenges, for which there is no condition similar to the last duty first. There may therefore be several **P**-moves which **O** can challenge in an infinite number of ways (in accordance with the Posit-Substitution rule) at a given stage. Thus we partition the set $Sc(t)$ of nodes in $S(t)$ associated with an **O**-challenge depending on the **P**-move that is challenged.

[2]The first thorough study of the case of classical FOL has been developed by Clerbout (2014a, b).

Summing up, we consider the following partition of $S(t)$:

A $Sd(t) = \{n|n \in S(t)$ and n is associated with a defence$\}$
B For each **P**-node m from the root to t, the set
$Sc_m(t)$=$\{n|n \in S(t)$ and n is associated with a challenge against the **P** node $m\}$

Once again, we insist on the fact that since there are only finitely many nodes from the root to t there is only a finite number of sets Sc_m.

Thus, we have partitioned $S(t)$ in a finite collection of disjoint subsets, such that at least one of them is infinite—otherwise t would not be a critical node. Because our current aim is to get rid of infinite ramifications, we leave the finite subsets untouched. Notice that each subset $Sc_m(t)$ of challenges resulting from an application of a rule other than the Posit-Substitution rule is finite because it is the only rule in which the challenger has infinitely many choices for his challenges.

Suppose next that $Sd(t)$ is infinite. In this case *we keep exactly one of its members* in $\mathbf{S}^\star$ and *delete all the other members as well as the branches they generate*.

We can safely do this because as pointed out, all the members in $Sd(t)$ are defences in reaction to the same previous **P**-move: they result from an application of the rule $SR4.1$ for Resolution of Instructions. Therefore, all the moves associated with nodes in $Sd(t)$ are the same *modulo the play-object*. They are, so to say, substitutional variants of each other. The nodes are similar because for $s^\star$ to be a winning **P**-strategy, the Proponent must win no matter what **O** does: none of these variations changes anything in terms of **P**'s ability to win. Hence, in a way, nothing is lost by keeping only one member of $Sd(t)$.

The same reasoning applies to the infinite $Sc_i(t)$ sets since the reason they are infinite is basically the same: they represent an infinite number of possible choices of play-objects by **O**, though this time as challenges instead of defences. Hence, when some sets $Sc_i(t)$ are infinite we do the same and keep, *for each of these sets*, exactly one member and delete the others and the branch they generate.

Summing up, we partition the set of successors for every critical node in $\mathbf{S}^\star$ to obtain a finite number of disjoint subsets (some may be infinite). We leave the finite ones untouched and reduce the infinite ones to singletons. This operation generates a tree, called $\mathbf{S^f}$, with no critical node and in which infinite ramifications have been successfully eliminated without loosing important information.

Let us discuss once more the crucial point allowing us to build $\mathbf{S^f}$. Because $\mathbf{S}^\star$ is the extensive form of a winning **P**-strategy, we know that the Proponent wins in every branch and thus to some extent the play-object chosen by the Opponent for the instructions does not matter. This means that in the case of universal quantification posited by **P**, the Proponent has a method to successfully defend his posit no matter which play-object **O** chooses for $L^\forall(p) : A$ (where A is a set). This is exactly the natural-deduction description of the Introduction rule for universal quantification: it is harmless to keep only one of the possible choices by **O** because the existence of a winning **P**-strategy ensures that there is indeed a successful method for every possible choice by **O**.

3.1.2 Disregarding Formation Rules

The dialogical rules allow the players to enquire about the type of expressions and in particular to ask whether an expression is a proposition or not. This leads to plays using the Formation rules listed in Sect. 2.2. However, as we mentioned in the introduction, since we only cover in our study the fragment of CTT involving logically valid propositions, we will assume that the underlying sets have been well built and that all the expressions at stake are well typed. We will therefore ignore in $\mathbf{S^f}$ every branch in which a formation rule is applied: we simply remove these branches and call the obtained tree **S**.

3.1.3 Disregarding Irrelevant Variations in the Order of O-moves

A **P**-strategy must account for every possible way for **O** to play, and in particular it must deal with any order in which the Opponent might play her moves. It means that $\mathbf{S^\star}$ had branches differing from other branches only in the order in which **O** plays her moves, and therefore so does **S**. However, since we started with a winning **P**-strategy, we can select any particular order of **O**-moves without losing anything in terms of **P**'s victory. Indeed, by the very definition of a winning strategy, the order of **O**-moves does not influence the result: every branch of the tree extracted from **S** still represents a play won by **P**. Our next step will thus be to extract from **S** a tree representing only one order of **O**-moves: we are looking for a tree in which there are no two branches B_1 and B_2 identical to each other modulo the order of **O**-moves.

Nevertheless, using the intuitionistic development rule $SR1i$ requires some specific care while selecting a suitable play, for we do not want a play in which **O** loses because she played poorly. According to the rule $SR1i$, the Opponent can only answer the last **P**-challenge not yet answered. It tends therefore to be strategically safer for **O** to immediately defend (and be sure not to lose the chance of making that move later on) and delay possible moves involving counterattacks. When extracting a particular order of the Opponent's moves in **S** we shall thus select a tree $\mathbf{S^O}$ such that, in any branch, every **P**-challenge is immediately followed by the **O**-defence.[3] By doing so we explicitly get rid of the cases in which **O** loses only because she poorly chose the order of her moves: victories due to **O** playing poorly are not enough to conclude that **P** can always win, obviously not making good starting points for proofs either. This is why one starts with **P**'s strategies, accounting for every way for **O** to play to check the Proponent's ability to win, and ignore mere lucky victories when building a proof out of a winning strategy. Once we have removed all the redundant information for developing a demonstration, what remains is what we call the *core* **C** of $\mathbf{S^\star}$.

[3] Recall that the Opponent is always able to do so since unlike the Proponent she is not constrained by a rule like the Modified Formal Rule.

3.2 From the Core to a CTT Demonstration

The next step is to apply transformations to this core until we obtain a CTT demonstration. Rahman et al. (2009), on which we strongly rely in this section, simplifies the core further before translating it in a demonstration: question moves i.e., moves with '?', were removed from the core. This was justified since questions in standard dialogical games do not correspond to anything in standard natural deduction, mostly because the interactive process of question and answer makes no sense in natural deduction.

Although two-player interactions do not make sense in the CTT framework either, the situation is a bit different in our case. As a matter of fact some of the questions occurring in the branches of a dialogical core are important for the translation as a CTT demonstration, particularly in the case of questions resulting from instruction resolution/substitution (structural rules $SR4.1$-2) and various equality rules (e.g. set-equality).

Even so, not everything from the core **C** will be considered by the algorithm. The following will specifically be ignored:

- The players' identities.
- Choices of repetition ranks.
- Moves of the form "$sic(n)$" (see the discussion about atomic **P**-posits below).
- Questions played when applying the particle rules of Table 2.5.

Strictly speaking, only posits will be incorporated in the demonstration resulting from the translation algorithm. However, questions other than the ones from the rules of Table 2.5 will be useful in the translation process and as such will not be completely ignored. We give more details below.

Before delving into the algorithm, let us start with some terminology:

- Let π be some posit and p some play-object. We say that in **C** the move $\mathbf{X}\,!\,\pi$ is *case-dependent of* move $\mathbf{Y}\,!\,p : \varphi$ if **X** can defend π only after **Y** defended $p : \varphi$ and if this defence results from a choice by **Y**.
  Because of the second condition, it follows that φ must be of the form of a disjunction, an existential quantification or a subset separation.
- A *concession* is either:

  1. An **O**-posit conceded before **P** posits the thesis, or
  2. An **O**-posit played as a challenge against a **P** implication or a **P** universal (assumption).

- Let us say that, for a posit π occurring in the dialogical core **C**, the nodes *descending from* π are all the nodes which are related to π by a chain of applications of dialogical rules.
- When the dialogical core or the demonstration we are building splits, we speak of the left and right branches of the core (or demonstration). We may sometimes assign an order on the branches from left to right and speak of the first branch, second branch, etc.

3.2.1 Generalities

In a nutshell, what we take from Rahman et al. (2009) is the following correspondence within a **P**-strategy for the particle rules given in the previous chapter's Table 2.5: applications to **O**-posits correspond to Elimination rules, whereas Introduction rules correspond to applications to **P**-posits (provided we read these "bottom-up"). The exceptions to this general principle are

a. *Atomic posits made by the Proponent*. There are two cases for each of these posits: either it is challenged by **O**, or it is not. If it is not, it would be because the rule $SR3$ would have forbidden it, meaning that **O** had already posited the same atomic expression before, in which case the atomic **P**-posit results from the matching **O**-posit by applying rule $SR3$. On the other hand, if it *is* challenged, then again two possible cases arise: either the defence by **P** "$sic(n)$" (for some n) occurs in the core, or it does not. If it does occur in the core, the **P**-posit would also result from $SR3$. If it does *not* occur, then **O** must have posited $\bot$ in the same branch of the core, in which case the atomic **P**-posit results from applying the Elimination rule for $\bot$.
   Let us briefly explain the case involving the Elimination rule for $\bot$. Assume that an atomic **P**-posit is challenged and that **P** cannot answer with the move "$sic(n)$". Suppose further that **O** does not posit $\bot$. Then there would be a way for **O** to win, namely making that challenge at the right moment. But this would be contradictory with our starting assumption of considering a winning **P**-strategy. Hence there must be something causing the Opponent to lose: the only way left is for her to have posited $\bot$.
b. **P**-*posits which are case-dependent of an* **O**-*posit*, as defined above. In such a case we say that the **P**-posit results from applying an Elimination rule to the move of which the **P**-posit is dependent.

Particle rule correspondences (Summary):

- **O**-posits:

Application of a dialogical rule to	Corresponds to
An **O** disjunction	Elimination rule for disjunction (may be related to a case-dependent **P**-posit)
An **O** conjunction	Elimination rule for conjunction
An **O** existential	Elimination rule for existential (may be related to a case-dependent **P**-posit)
An **O** subset separation	Elimination rule for subset separation (may be related to a case-dependent **P**-posit)
An **O** implication	Elimination rule for implication
An **O** universal	Elimination rule for universal

- **P**-posits:
  - If they are atomic, they occur in the core resulting from the application either of rule $SR3$, or of absurdum Elimination.
  - If they are case-dependent of an **O**-posit then they result from applying an Elimination rule for disjunction, existential quantification or subset separation, depending on the **O**-posit.
  - Otherwise:

Application of a dialogical rule (read "bottom-up") to	Corresponds to
A **P** disjunction	Introduction rule for disjunction
A **P** conjunction	Introduction rule for conjunction
A **P** existential	Introduction rule for existential quantification
A **P** subset separation	Introduction rule for subset separation
A **P** implication	Introduction rule for implication
A **P** universal	Introduction rule for universal quantification

Resolution and substitution of instructions (rules $SR4.1$ and $SR4.2$). One of the cases for which it is important that the algorithm does not ignore the question mark '?' is the application of the structural rules $SR4.1$ and $SR4.2$ allowing to substitute dialogical instructions by play-objects. The algorithm takes them into account through the following operation which we shall refer as *Establishing Substitutions for Instructions* (ESI):

Assume that some instruction occurs in move number n. Scan the core **C**: if move n is challenged by a question of the form "$I/?$" for some instruction I and in virtue of rule $SR4.1$, then scan **C** in search for the defence. Same procedure if move n is challenged by a question of the form "$?I/p_i$" in virtue of rule $SR4.2$. The ESI operation consists in implementing the substitution while placing move n in the demonstration under way. That is, we place the expression $\pi(I/p')$ in the demonstration, where π is move n and p' is the play-object to be substituted for instruction I. Assume this is done at stage s_i of the translation of **C** into a demonstration. In practice it is often more convenient to simply write the result of the substitution from stage $s_i + 1$ on. For an illustration, see in particular the detailed example of the Axiom of Choice in Chap. 4.

Other rules. Apart from the particle rules of Table 2.5, the remaining rules will not be interpreted as Introduction or Elimination rules. These are the Equality and Substitution rules of Chap. 2, which will be related in the same way to their direct CTT counterpart no matter whether there are applied to **P**-moves or to **O**-moves.

3.2.2 The Algorithm

The procedure we describe hereafter can be seen as a rearrangement of (some of) the nodes in **C** eventually producing a CTT demonstration. However, we assume that we have an unmodified "copy" of **C** to which we can refer to while the procedure goes on. The last stage (**D**) of the procedure requires some explanation which we give after the procedure.

A—Initial stage. Let π be the thesis posited by **P** and $\gamma_1, \ldots, \gamma_n$ be the initial concessions, if any, posited by **O**. Place π as the conclusion of the demonstration under way and $\gamma_1, \ldots, \gamma_n$ as its uppermost premisses:

$$\begin{array}{c} \gamma_1 \quad \gamma_2 \quad \cdots \quad \gamma_n \\ \vdots \\ \pi \end{array}$$

Apply the Establishing Substitution for Instructions (ESI) operation described in the previous section and go to B.[4]

B. Consider the lowest expression π_i added in the branch of the demonstration being constructed.[5] Take the move corresponding to this expression in **C** and scan the core in order to know what dialogical rule has been applied to it and to identify the relevant challenge and defence. Then:

B.1. If the relevant challenge and defence have already been accounted for in the branch being constructed, then go to C. Otherwise go to B.2.

B.2. If the defence is "$sic(n)$", then apply stage B.2.a. Otherwise, apply stage B.2.b.

B.2.a. The move corresponding to π_i in **C** is an atomic **P**-posit. Draw an inference line above it and label it as application of rule $SR3$. If the corresponding **O**-posit has already been accounted for in the current branch, then rearrange the branch so that it is placed as the premiss of the application of the rule. Go back to B.

B.2.b. Draw an inference line above π_i and label it with the name of the relevant rule. Place the defence—and, if relevant, the challenge[6]—as the premiss (or premisses) of the application of the rule, according to the following conventions:

- In the case of the Introduction rule for material implication, negation or universal quantification, the defence is the immediate premiss and the challenge is placed upwards as an assumption such that: (i) the defence depends on that assumption, (ii) the assumption is numbered and marked as discharged at the inference step and (iii) this assumption is still in the scope of previously placed assumptions such as the premisses of the demonstration placed in stage A.

[4]If there is no room for an application of the ESI operation (if no instruction occurs in the expressions considered), the algorithm simply ignores this step and proceeds.

[5]Ramifications in the demonstration may have been produced during the progression of the translation, as described in B.2.b. If there are no ramifications, then we are still in the main branch of the demonstration.

[6]When the challenge is not a question, that is in the particle rules for material implication, negation and universal quantification.

- In the other cases having multiple premisses, the premisses do not depend on one another and are placed on the same level, each opening a new branch in the demonstration. In such cases all the premisses placed at some previous step in the translation must be copied and pasted for each newly opened branch.

Apply the ESI operation to the newly added expressions. If relevant, move to the first (starting from the left) newly opened branch and go back to B.

C. If the situation is the one of B.1 and no new expression has been added to the branch being constructed, then:

C.1. Perform any rearrangement required to match the notational convention of CTT, and go to C.2.

C.2. If the branch does not feature applications of rule $SR3$, then go to C.3. Otherwise for each application of $SR3$ in the branch, remove its conclusion[7] and the associated inference line. Go to C.3.

C.3. Move to the next branch to which stages C.1 and C.2 have not been applied and go back to B. If there is no such branch left then go to D.

D. Going from the top to the bottom, replace in the demonstration at hand the dialogical play-objects with CTT proof-objects in accordance with the CTT rules. Then stop the procedure.

A note on steps C.1-3. We have designed the algorithm so that the branches in the demonstration under construction are dealt with sequentially. However, it is possible to treat them all at the same time in parallel.

Explanations on the final stage. The concluding stage D is necessary because, as we pointed out several times, dialogical play-objects differ from CTT proof-objects. If we attempt to build up the demonstration with play-objects only, the procedure would not yield a correct CTT demonstration even if we assume that it works adequately on other respects.[8] There would be a fundamental difference, since the operations defined on play-objects by dialogical rules are not the same as the operations defined on proof-objects by CTT rules. Let us first describe in more detail how the replacement of play-objects by proof-objects is carried out. After that, we will argue that such a replacement is justified in spite of the differences between play-objects and proof-objects.

We start by freely replacing play-objects occurring in the leafs of the demonstration at hand. At this point it is simply a matter of rewriting since starting a CTT demonstration only requires suitable proof-objects for each of the relevant assumptions: which proof-objects have been chosen does not really matter in this context as long as they are of the required type. While building the demonstration out of the dialogical core we have identified the rules which are applied. We can thus refer to the CTT rules of Chap. 1 to know how new proof-objects are obtained depending on the rule applied, and substitute them for the dialogical play-objects accordingly.

[7] As previously mentioned, this would be a repetition of the premiss of the rule: the conclusion corresponds to an atomic **P**-posit which from the point of view of CTT is redundant since the same expression coming from the corresponding **O**-posit already occurs.

[8] We address this part of the question in Sect. 3.3.

It is helpful to consider the history of substitution of play-objects for instructions and how they were implemented in the transformation of the core into a demonstration: the syntax of embedded instructions mirrors the sequence of applications of rules and can thus bring some insight on the origins of the associated play-objects.[9]

Replacing play-objects by proof-objects has two prerequisites:
First, the construction of the derivation must observe the rules of CTT and in particular the order according to which some proof-objects are obtained from other proof-objects occurring previously in the derivation. This means that play-objects occurring in the conclusion of rule applications must be obtained from previous play-objects according to a dialogical rule. But applications of dialogical rules are not necessarily similar to applications of CTT rules in that respect, especially in the case of **P**-applications which must be read bottom-up when translated in the CTT framework. This is the point where the history of the substitution of instructions is helpful since the syntax of instructions enables us to keep track of how a given complex play-object results from simpler ones. The adequacy of the algorithm, which we will address in the next section of the present chapter, will then guarantee that the derivation proceeds according to the rules of CTT.

Second, the replacement is carried out while transforming the core of a *winning* **P**-strategy. Only then can play-objects be related not only to the use (meaning) of expressions, but also to their justification. This cannot be achieved while considering single plays—nor non-winning strategies, obviously. Consider for example the case of **P**-conjunctions. In general, single plays cannot provide a way to check if a conjunction is justified: this would require **P** to win the play for the two conjuncts, only there is always at least one case for which this cannot be done within a single play. Such is the case if the Opponent cannot ask for both conjuncts in the same play because of her repetition rank, and we have precisely started the extraction of the core by narrowing the focus to the case in which her rank is 1.

Another clear example is the case of material implication and universal quantification: it is only when we consider every possible way for **O** to play—i.e. a **P**-strategy—that we can check if there is a way for **P** to justify the second part for every play-object chosen by **O** for the first part or not. Actually, in these cases a winning **P**-strategy literally displays the procedure by which the Proponent chooses the play-object for the second part of the expression at stake depending on the play-object chosen by the Opponent for the first part. This corresponds to the general description of proof-objects for material implications and universally quantified formulas in CTT: a method which, given a proof-object for the first part, yields a proof-object for the second part. The dialogical interpretation of this functional dependence amounts to a choice dependence.

As indicated, we will study in the next chapter how our algorithm is applied to the case of the Axiom of Choice which strikingly displays the force and fruitfulness of the CTT approach. We will discuss and explain at the same time the dialogical take

[9]For an application of this method, see the detailed discussion of the proof of the Axiom of Choice in next chapter.

on this axiom. However, before doing so the algorithm must be proven adequate. This brings us to the last section of this chapter.

3.3 Adequacy of the Translation Algorithm

We must ensure that the algorithm is adequate: given the core of a winning **P**-strategy it must always yield a CTT demonstration. Let us first describe the general idea behind the demonstration.

The translation procedure ultimately consists in rearranging the nodes of the original dialogical core **C**. We must ensure that the reordering results in a derivation which complies with the CTT rules. We noticed that during this reordering, the procedure introduces what we may call "gaps" which we have marked with vertical dots. Take for example the first step of such a transformation procedure. In this step the thesis of the core gives the conclusion of the demonstration and the concessions give the assumptions, though we still do not know at this point of the process what corresponds to the steps between the assumptions and the conclusion. Accordingly, we start by simply linking the assumptions and the conclusion with vertical dots. The idea behind the adequacy of the algorithm is that all these gaps will eventually be filled and that it will be done in a way which observes the CTT rules.

The last part of this statement is easily checked. Let us assume that all the gaps are indeed removed. Then we can easily see that the resulting derivation is such that every rule applied in it is a CTT rule. We have indeed associated every application of a dialogical rule to a CTT rule, with three notable exceptions: the Modified Formal Rule $SR3$, the Particle rule for atomic expressions and the rules for Resolution and Substitution for Instructions $SR4.1$ and $SR4.2$. But applications of these three rules will eventually be removed too:

(i) At the last stage of the algorithm, when play-objects are replaced by proof-objects, applications of the rules $SR4.1$ and $SR4.2$ regarding instructions will eventually be removed.
(ii) Applications of the particle rule for atomic expressions are ignored since they are dealt with in terms of the Elimination rule for $\perp$ or in terms of the modified formal rule $SR3$.
(iii) Applications of the rule $SR3$ are eventually removed together with the redundant atomic expressions they introduced.

So far so good—though the critical task of checking that the CTT rules are *properly* applied still remains. This process must show the important fact that following the algorithm will eventually remove the gaps, as it was assumed above. In order to ground this assumption let us temporarily consider an extension of the CTT calculus which includes the rules $SR3$ and $SR4.1$-2 as well as a new rule called *Gap* allowing either to link (with the help of vertical dots) two nodes of the demonstration without a dialogical rule explaining such a link, or to introduce an expression as the last step of a sequence of vertical dots. We will show that when following the algorithm, each of the applications of the rule *Gap* will be replaced by applications of a suitable CTT rule or by applications of one of $SR3$, $SR4.1$-2. We will then simply need to show that when no dialogical rule is applied to the corresponding node from **C**, the expression will not introduce additional gaps: the rearranging in the stage C.1 of the algorithm is harmless. Once we have reached this point, and after all applications of $SR3$ and $SR4.1$-2 have been removed, we are assured to have a proper CTT demonstration.

Accordingly, let us show first that the gaps introduced during the process of building the CTT demonstration are temporary and will be progressively removed bottom-up:

Proposition 3.1 *For any stage of the translation procedure, there is a corresponding node in the original dialogical core* **C** *for every expression resulting from a gap.*

Proof The proof is an easy but fastidious induction which also establishes that newly introduced gaps at a given stage of the translation have the "right shape", so that they will be filled by a proper application of a rule later on. The *base case* is trivial: the initial stage A of the algorithm stipulates that the first expression resulting from an application of the rule "Gap" is the thesis, which is obviously a node in **C** to which a dialogical rule is applied.

Inductive Hypothesis. Assume that the Proposition holds for every application of the rule "Gap" up to this step in the translation procedure, say after n steps. We show that the Proposition holds for the gaps introduced at step $n+1$ and that they have the correct "shape" in relation to the development of a CTT demonstration. This is done by cases, depending on the form of the last expression introduced at this point. For simplicity and brevity we only spell out two cases:

- The associated node in **C** is a **P** disjunction $p : A \vee B$ which is not case-dependent, and the fragment of the derivation at stake at step n is:

$$\begin{array}{c} \vdots \\ p : A \vee B \end{array}$$

then according to the algorithm, the result at step $n+1$ is:

$$\dfrac{\begin{array}{c} \vdots \\ L^{\vee}(p) : A \end{array}}{p : A \vee B}\ \vee I$$

We next recall that we must have **O** challenging the disjunction at some place in the core: if there is a **P**-move in **C** which **O** does not challenge—though he could—then the core contains branches which do not represent terminal plays. However this is not possible since we have assumed **C** to be the core of a winning **P**-strategy. For the same reason, the core must feature the successful defence by the Proponent of one of the disjuncts, say A. Thus the newly added expression filling up the dots introduced by *Gap* does indeed correspond to a node in **C**.

- The associated node in **C** is a **P** conjunction $p : A \wedge B$ which is not case-dependent. After step n we then have:

$$\begin{array}{c}\vdots\\ p : A \wedge B\end{array}$$

so that according to the algorithm the result at step $n + 1$ is:

$$\dfrac{\begin{array}{c}\vdots\\ L^{\wedge}(p) : A\end{array} \quad \begin{array}{c}\vdots\\ R^{\wedge}(p) : B\end{array}}{p : A \wedge B} \wedge I$$

Just like in the previous case, we must have **O** challenging the **P** conjunction at some place in the core—otherwise **C** contains non-terminal plays and we have a contradiction—resulting in a ramification in which each branch contains the posit by **P** of one of the conjuncts. Expansion of the demonstration thus follows the CTT rule and the new expressions filling up the dots introduced by *Gap* correspond to these nodes in **C**.

The construction of the demonstration thus proceeds by progressively filling up the temporary gaps until it reaches a stage at which no further gap is introduced. Except for the initial assumptions of the demonstration,[10] the cases in which no gaps are introduced are reduced to cases of atomic expressions. But these come either from an Elimination rule for absurdum or from $SR3$, that is, precisely the cases for which the premisses must already have been processed.

Summing up, the demonstration by induction of Proposition 3.1 shows that the algorithm builds a derivation by introducing temporary gaps and then progressively filling them up until no further gap occurs. Moreover, this construction has been developed in such a way that the derivation complies with the proceedings of what we have called the extended CTT calculus (which includes $SR3$ and $SR4.1$-2 in addition to the usual CTT rules).

Finally, as we have pointed out at the beginning of this section, the applications of the rules that do not strictly pertain to Constructive Type Theory are removed to guarantee that only CTT rules are applied in the resulting derivation. From all this together we have the following corollary:

Corollary 3.1 *Let* **C** *be the core of a winning* **P***-strategy in the game for* $p : \varphi$ *under initial concessions* $\gamma_1, \ldots, \gamma_n$. *The result of applying the translation algorithm to* **C** *is a CTT demonstration of* φ *under the hypotheses* $\gamma_1, \ldots, \gamma_n$.

This concludes the chapter. For the demonstration of the equivalence between dialogical games and CTT, we need to consider the converse direction, namely from a CTT demonstration to a winning **P**-strategy. Before doing so, however, we will

[10]Which correspond in the core to the initial concessions of **O**.

introduce an application of this chapter's algorithm with the example of the Axiom of Choice, considering both its intensional and extensional versions. By doing so we will also be able to introduce and explain the dialogical take on this axiom. Applying the algorithm will also minutely illustrate in detail the procedure of "filling the gaps" described by Proposition 3.1 above.

References

Clerbout, N. 2014a. First-order dialogical games and tableaux. *Journal of Philosophical Logic* 43(4): 785–801. [Published first online 2013. doi:10.1007/s10992-013-9289-z].

Clerbout, N. 2014b. *La Sémantique Dialogique. Notions Fondamentales et Éléments de Metathéorie*. London: College Publications.

Rahman, S., N. Clerbout, and L. Keiff. 2009. On dialogues and natural deduction. In *Acts of Knowledge: History, Philosophy and Logic: Essays Dedicated to Göran Sundholm*, ed. G. Primiero and S. Rahman, 301–336. London: College Publications.

Chapter 4
The Dialogical Take on the Axiom of Choice, and Its Translation into CTT

To illustrate the translation procedure described in the previous chapter, here is what is known as the intensional formulation of the Axiom of Choice. The formulation

$$(\forall x : A)(\exists y : B(x))C(x, y) \rightarrow (\exists f : (\forall x : A)B(x))(\forall x : A)C(x, f(x))$$

as well as the CTT demonstration have been described in Martin-Löf (1984, pp. 50–51).

Besides a simple illustration, there is a good argument for using the dialogical approach to meaning which supports a game theoretical reading of the intensional formulation of the Axiom of Choice described above. In fact, to the best of our knowledge, Hintikka (1996a) was the first to provide a game-theoretical reading of the axiom. But, as discussed by Jovanovic,[1] Hintikka's interpretation is quite debatable.[2] Nevertheless there is a valuable point in a game theoretical reading which has been stressed by recent work on dialogical approaches to CTT by Rahman and Clerbout (2013). That is, if meaning is conceived as being constituted during interaction, then all of the actions involved in the constitution of the meaning of an expression should be made explicit. They should all be part of the object language. This perspective is rooted in Wittgenstein's *UnHintergehbarkeit der Sprache*. Besides, language-games are purported to accomplish the task of displaying this "internalist feature of meaning". One of the main insights of Kuno Lorenz' interpretation of the

[1] Jovanovic [forthcoming].

[2] If we express at the object language level the domain and codomain of the function, Hintikka's formulation of the Axiom of Choice coincides with the intensional one. However, as Jovanovic discusses, Hintikka tries to render its meaning via a non-constructive semantics based on IF-logic.

N. Clerbout and S. Rahman, *Linking Game-Theoretical Approaches with Constructive Type Theory*, SpringerBriefs in Philosophy,
DOI 10.1007/978-3-319-19063-1_4

relation between the so-called first and second Wittgenstein is based on a thorough criticism of the metalogical approach to meaning (Lorenz 1970, pp. 74–79).[3]

Recalling how Hintikka (1996b) extended the way van Heijenoort differentiates a language as the universal medium from a language as a calculus, we see that the dialogical approach shares tenets from both conceptions. On the one hand, the dialogical approach shares with universalists the view that we cannot place ourselves outside our language. On the other, it shares with the anti-universalists the view that we can methodically reconstruct a given linguistic practice which is complex out of the interaction of simple ones. The CTT approach to meaning in general and to the Axiom of Choice in particular is thus very natural to game theoretical approaches in which (standard) metalogical features are explicitly displayed at the object language level. In fact, from the dialogical point of view, the actions constituting the meaning of logical constants, such as choices, are a key element of its fully fledged (local) semantics.

4.1 The Dialogical Take on the Axiom of Choice

Since the work of Martin-Löf, the intensional formulation of the Axiom of Choice is evident in the sense that it is logically valid. As pointed out by Bell (2009, p. 206) its logical validity entitles us to call it an axiom rather than a postulate (as in its classical or extensional version, which is not valid).[4] Before exhaustively developing the winning strategy for the intensional Axiom of Choice let us formulate the idea behind the dialogical approach by mirroring Martin-Löf's own presentation of the informal constructive demonstration of it (1984, p. 50).[5]

From the dialogical point of view the point is that **P** can copy-cat **O**'s choice about y in the antecedent for $f(x)$ when defending the consequent. This is because both are equal objects of type $B(x)$, for any $x : A$. A winning strategy for the implication thus

[3]Similar criticism has been raised by Sundholm (1997, 2001) who points out that the standard model-theoretical approaches to meaning turn semantics into a meta-mathematical formal object in which syntax is linked to semantics by the assignation of truth values to uninterpreted strings of signs (formulae). Language does not express content any more but rather is conceived as a system of signs speaking about the world—provided a suitable metalogical link between signs and world has been fixed.

[4]See Sect. 4.3.

[5]See also Bell (2009, pp. 203–204) who uses the notation of Tait (1994). Tait's functions π and π' are very close to our left and right instructions—though we differentiate instructions for each logical constant by adding a superscript to identify them. However, we do not have explicitly Tait's function σ, though the result of the substitution of an instruction with a pair of embedded instructions—what we call its resolution—will yield the pair of this function's components.

simply follows from the meaning of the antecedent. The embedding of quantifiers results in an interaction of choices (of play-objects) which generate dependences on which the Proponent can rely for choices of his own leading him to victory. In more details:

- Let us assume that the Opponent launches an attack on the implication and accordingly posits its antecedent—the play-object for the antecedent being $L^{\rightarrow}(p)$. Let us further assume that with her challenge **O** resolves the instruction $L^{\rightarrow}(p)$, by choosing v.
- Then for any $x : A$ chosen by **P**, there must be a play-object $R^{\forall}(v)$ for the right component of the antecedent.
- However, $R^{\forall}(v)$ is a play-object for an existential and is thus composed of two play-objects such that the first one $L^{\exists}(R^{\forall}(v))$ is of type $B(x)$ for any $x : A$ and the second one $R^{\exists}(R^{\forall}(v))$ is, for any $x : A$, of type $C(x, L^{\exists}(R^{\forall}(v)))$.
- The Proponent defends the implication by positing the consequent—with $R^{\rightarrow}(p)$ as play-object. Since the consequent is an existential, $R^{\rightarrow}(p)$ is composed of two play-objects. The key point is then to let **P** choose precisely v as the first component of $R^{\rightarrow}(p)$. This component is, for any $x : A$, of type $B(x)$. Thus, the first component of the play-object for the existential in the consequent is of the same type as the left component of the existential in the antecedent. They are not only of the same type: since **P** copies (when defending the existential) the choice of **O**, namely v, we are entitled to say that the left component of the play-object for the existential in the consequent is exactly the same in $B(x)$ as the left component of the existential occurring in the antecedent—i.e., $y = v(x) : B(x)$.
- Since, in the antecedent, the y of $C(x, y)$ is of type $B(x)$ for any $x : A$, and since, as already mentioned, y is equal to $v(x)$ in $B(x)$, then it follows that $C(x, y)$ in the antecedent is, for any $x : A$, intensionally equal to $C(x, v(x))$ in the type set. More generally, and independently of **O**'s particular choices for the play-object for the antecedent and of x, **P** can establish by this copy-cat strategy that $C(x, y)$ and $C(x, f(x))$ are two equal sets (for any $x : A$ and for $y : B(x)$).

It follows from the last two steps that **P** can copy-cat the play-object for the antecedent into the play-object for the consequent. This copy-cat is the basic idea underlying the existence of a winning strategy for the Proponent, and will eventually allow replacing the play-objects by proof-objects.

Let us come back to our algorithm: since we focus here on the procedure by which we transform the core of a winning **P**-strategy into a CTT demonstration, we will assume that the simplification process extracting that core from the strategy has already been carried out. Accordingly, we start with a given core, described in Sect. 4.1.2 below. However, for the sake of clarity, we will start by spelling out the two plays displayed in the core, by means of the standard table notation of dialogical logic. This should also provide an insight on the interactions involved in the dialogical take on the Axiom of Choice. Recall that some initial concessions account for the part related to the formation of expressions. It is assumed to be unproblematic since we are dealing with the valid fragment of CTT.

4.1.1 Two Plays on the Axiom of Choice

We formally develop below the two plays constituting the core of the strategy. They are triggered by the Opponent's options at move 9 when challenging the existential posited by the Proponent at move 8. Since **O**'s repetition rank is 1, she cannot perform both challenges within a single play, hence the distinction between the two following plays. The first play (Table 4.1) corresponds in the demonstration to the Introduction of the universal in the consequent, under the assumption of the antecedent. The second play (Table 4.2) develops all the points of the informal demonstration described above.

Description of the first play:

- *Move* 3: Once the initial concessions and the thesis have been posited, and once the players have chosen their repetition rank, **O** launches an attack on material implication.
- *Move* 4: **P** launches a counterattack and asks for the play-object corresponding to $L^{\rightarrow}(p)$.
- *Moves* 5, 6: **O** responds to the challenge of 4. **P** posits the right side of the material implication.
- *Moves* 7, 8: **O** asks for the play-object corresponding to $R^{\rightarrow}(p)$. **P** responds to the challenge by choosing the pair (v, r) where v is the play-object chosen to substitute for the variable f, and r is the play-object for the right side of the existential.
- *Move* 9: Here **O** can choose between asking for the left side of the existential and asking for the right side. This first play describes how the game proceeds when **O** makes the left choice.
- *Moves* 10-26: These moves follow from a straightforward application of the dialogical rules. Move 26 is an answer to move 13, which **P** makes after he gathered enough information to be able to apply the characteristic copy-cat method imposed by the structural rule *SR*3.
- *Moves* 27, 28: **O** asks for the play-object corresponding to the instruction posited by **P** at move 26 and **P** answers and wins by applying copy-cat to **O**'s move 25. Notice that move 28 is not a case of function substitution: it is simply the resolution of an instruction.

Table 4.1 First play for the axiom of choice

	O			P	
$H1$	$C(x, y) : set\ (x : A,\ y : B(x))$				
$H2$	$B(x) : set\ (x : A)$				
$H3$	$A : set$				
				$p : (\forall x : A)(\exists y : B(x))C(x, y) \to (\exists f : (\forall x : A)B(x))(\forall x : A)C(x, f(x))$	0
1	$\mathtt{n} := 1$			$\mathtt{m} := 2$	2
3	$L^{\to}(p) : (\forall x : A)(\exists y : B(x))C(x, y)$	(0)		$R^{\to}(p) : (\exists f : (\forall x : A)B(x))(\forall x : A)C(x, f(x))$	6
5	$v : (\forall x : A)(\exists y : B(x))C(x, y)$		(3)	$L^{\to}(p)/?$	4
7	$R^{\to}(p)/?$	(6)		$(v, r) : (\exists f : (\forall x : A)B(x))(\forall x : A)C(x, f(x))$	8
9	$?_L$	(8)		$L^{\exists}(v, r) : (\forall x : A)B(x)$	10
11	$L^{\exists}(v, r)/?$	(10)		$v : (\forall x : A)B(x)$	12
13	$L^{\forall}(v) : A$	(12)		$R^{\forall}(v) : B(w)$	26
15	$w : A$		(13)	$L^{\forall}(v)/?$	14
19	$R^{\forall}(v) : (\exists y : B(w))C(w, y)$		(5)	$L^{\forall}(v) : A$	16
17	$L^{\forall}(v)/?$	(16)		$w : A$	18
21	$(t_1, t_2) : (\exists y : B(w))C(w, y)$		(19)	$R^{\forall}(v)/?$	20
23	$L^{\exists}(t_1, t_2) : B(w)$		(21)	$?_L$	22
25	$t_1 : B(w)$		(23)	$L^{\exists}(t_1, t_2)/?$	24
27	$R^{\forall}(v)/?$	(26)		$t_1 : B(w)$	28

The opponent's move 9 asks for the left play-object for the existential quantification on f

Table 4.2 Second play for the axiom of choice

	O			P	
$H1$	$C(x, y) : set\ (x : A,\ y : B(x))$				
$H2$	$B(x) : set\ (x : A)$				
$H3$	$A : set$				
				$p : (\forall x : A)(\exists y : B(x))C(x, y) \rightarrow (\exists f : (\forall x : A)B(x))(\forall x : A)C(x, f(x))$	0
1	$\mathtt{n} := 1$			$\mathtt{m} := 2$	2
3	$L^{\rightarrow}(p) : (\forall x : A)(\exists y : B(x))C(x, y)$	(0)		$R^{\rightarrow}(p) : (\exists f : (\forall x : A)B(x))(\forall x : A)C(x, f(x))$	6
5	$v : (\forall x : A)(\exists y : B(x))C(x, y)$		(3)	$L^{\rightarrow}(p)/?$	4
7	$R^{\rightarrow}(p)/?$	(6)		$(v, r) : (\exists f : (\forall x : A)B(x))(\forall x : A)C(x, f(x))$	8
9	$?_R$	(8)		$R^{\exists}(v, r) : (\forall x : A)C(x, L^{\exists}(v, r)(x))$	10
11	$L^{\exists}(v, r)/?$	(10)		$R^{\exists}(v, r) : (\forall x : A)C(x, v(x))$	12
13	$R^{\exists}(v, r)/?$	(12)		$r : (\forall x : A)C(x, v(x))$	14
15	$L^{\forall}(r) : A$	(14)		$R^{\forall}(r) : C(w, v(w))$	32
17	$w : A$		(15)	$L^{\forall}(r)/?$	16
21	$R^{\forall}(v) : (\exists y : B(w))C(w, y)$		(5)	$L^{\forall}(v) : A$	18
19	$L^{\forall}(v)/?$	(18)		$w : A$	20
23	$(t_1, t_2) : (\exists y : B(w))C(w, y)$		(21)	$R^{\forall}(v)/?$	22
25	$L^{\exists}(t_1, t_2) : B(w)$		(23)	$?_L$	24
27	$t_1 : B(w)$		(25)	$L^{\exists}(t_1, t_2)/?$	26
29	$R^{\exists}(t_1, t_2) : C(w, t_1)$		(23)	$?_R$	28
31	$t_2 : C(w, t_1)$		(29)	$R^{\exists}(t_1, t_2)/?$	30
33	$R^{\forall}(r)/?$	(32)		$t_2 : C(w, v(w))$	34
35	$v(w)/?_{<=>}$	(34)		$t_2 : C(w, t_1) < C(w, t_1) = C(w, t_1/v(w)) : set >$	42
41	$C(w, t_1) = C(w, t_1/v(w)) : set$		$(H1)$	$v(w) = t_1 : B(w)$	36
37	$v(w) = t_1 : B(w)?$	(36)		$sic(39)$	40
39	$v(w) = t_1 : B(w)$		(5, 20-21, 27)	$?_{\Pi-eq}$	38

The opponent's move 9 asks for the right play-object for the existential quantification on f

Description of the second play:

- *Move* 9: The play differs from the first one when **O** chooses to ask for the right side of the existential posited by **P** at move 8.
- *Moves* 10-34: The moves follow from a straightforward application of the dialogical rules.
- *Move* 35: **O** asks the Proponent to perform a function-substitution for $v(w)$.
- *Move* 36: The Proponent's aim is to substitute it with t_1 so that he can take advantage of the Opponent's posit of $t_1 : C(w, t_1)$ at move 31. But in order to do so, he must establish that $C(w, t_1)$ and $C(w, t_1/v(w))$ are equal sets. For that purpose he launches a counterattack, challenging the Opponent's first initial concession $H1$ by positing that $v(w)$ and t_1 are equal in $B(w)$.
- *Move* 37: **O** launches a counterattack on **P**'s posit in move 36. Indeed she has not posited this equality herself, therefore she can challenge it instead of answering directly to **P**'s challenge 36.
- *Move* 38: To be able to answer the Opponent's last challenge, the Proponent must make her posit $v(w) = t_1 : B(w)$. To this end, he applies the Π-Equality rule.[6] Indeed, enough information has been gathered during the play for him to do so. **O** has indeed conceded in move 21 that, under the assumption that $w : A$, there is a y in $B(w)$ such that $C(w, y)$. The Proponent, based on **O**'s own move 17, has granted the assumption with his move 20. Hence, from **O**'s move 5, this y is the result of applying v to $w : A$. Finally, **O** has chosen t_1 to be this y in $B(w)$ with her move 27.
- *Move* 39: **O** has thus no other choice but to concede that $v(w) = t_1 : B(w)$.
- *Move* 40: This allows **P** to answer **O**'s challenge 37 with "*sic* (39)".
- *Move* 41: Now that **P** has successfully responded to her counterattack, **O** has no choice but to answer **P**'s challenge 36 and posit the set equality $C(w, t_1) = C(w, t_1/v(w)) : set$.
- *Move* 42: **P** is now finally in position to answer **O**'s challenge 35: he can posit that $t_2 : C(w, t_1)$ under the proviso that $C(w, t_1) = C(w, t_1/v(w)) : set$.
  There is nothing **O** can do now: she can challenge neither the posit nor the proviso, because she granted them herself with her moves 27 and 41. The Proponent wins this second play.

4.1.2 The Core of a Winning **P**-*strategy*

As previously said, the core we will work with consists in the two plays we have just described and explained. The core is given in Fig. 4.1, p. 60. In order to identify the dialogical source of each move we use [?*n*] to indicate the attacked line and [*m*] to indicate the challenge of player **X** triggering the posited defence of player **Y**.

[6]See the rule in Sect. 2.3, Chap. 2.

$H1.\mathbf{O}\text{-}C(x,y):set\ (x:A,y:B(x))$
$H2.\mathbf{O}\text{-}B(x):set\ (x:A)$
$H3.\mathbf{O}\text{-}A:set$
$0.\mathbf{P}\text{-}p:(\forall x:A)(\exists y:B(x))C(x,y)\rightarrow(\exists f:(\forall x:A)B(x))(\forall x:A)C(x,f(x))$
$1.\mathbf{O}\text{-}\mathtt{n}:=1$
$2.\mathbf{P}\text{-}\mathtt{m}:=2$
$3.\mathbf{O}\text{-}L^{\rightarrow}(p):(\forall x:A)(\exists y:B(x))C(x,y)\text{-}[?0]$
$4.\mathbf{P}\text{-}L^{\rightarrow}(p)/?\text{-}[?3]$
$5.\mathbf{O}\text{-}v:(\forall x:A)(\exists y:B(x))C(x,y)\text{-}[4]$
$6.\mathbf{P}\text{-}R^{\rightarrow}(p):(\exists f:(\forall x:A)B(x))(\forall x:A)C(x,f(x))\text{-}[3]$
$7.\mathbf{O}\text{-}R^{\rightarrow}(p)/?\text{-}[?6]$
$8.\mathbf{P}\text{-}(v,r):(\exists f:(\forall x:A)B(x))(\forall x:A)C(x,f(x))\text{-}[7]$

Left branch:

$9.\mathbf{O}\text{-}?_L\text{-}[?8]$
$10.\mathbf{P}\text{-}L^{\exists}(v,r):(\forall x:A)B(x)\text{-}[9]$
$11.\mathbf{O}\text{-}L^{\exists}(v,r)/?\text{-}[?10]$
$12.\mathbf{P}\text{-}v:(\forall x:A)B(x)\text{-}[11]$
$13.\mathbf{O}\text{-}L^{\forall}(v):A\text{-}[?12]$
$14.\mathbf{P}\text{-}L^{\forall}(v)/?\text{-}[?13]$
$15.\mathbf{O}\text{-}w:A\text{-}[14]$
$16.\mathbf{P}\text{-}L^{\forall}(v):A\text{-}[?5]$
$17.\mathbf{O}\text{-}L^{\forall}(v)/?\text{-}[?16]$
$18.\mathbf{P}\text{-}w:A\text{-}[17]$
$19.\mathbf{O}\text{-}R^{\forall}(v):(\exists y:B(w))C(w,y)\text{-}[16]$
$20.\mathbf{P}\text{-}R^{\forall}(v)/?\text{-}[?19]$
$21.\mathbf{O}\text{-}(t_1,t_2):(\exists y:B(w))C(w,y)\text{-}[20]$
$22.\mathbf{P}\text{-}?_L\text{-}[?21]$
$23.\mathbf{O}\text{-}L^{\exists}(t_1,t_2):B(w)\text{-}[22]$
$24.\mathbf{P}\text{-}L^{\exists}(t_1,t_2)/?\text{-}[?23]$
$25.\mathbf{O}\text{-}t_1:B(w)\text{-}[24]$
$26.\mathbf{P}\text{-}R^{\forall}(v):B(w)\text{-}[13]$
$27.\mathbf{O}\text{-}R^{\forall}(v)/?\text{-}[?26]$
$28.\mathbf{P}\text{-}t_1:B(w)\text{-}[27]$

Right branch:

$9.\mathbf{O}\text{-}?_R\text{-}[?8]$
$10.\mathbf{P}\text{-}R^{\exists}(v,r):(\forall x:A)C(x,L^{\exists}(v,r)(x))\text{-}[9]$
$11.\mathbf{O}\text{-}L^{\exists}(v,r)/?\text{-}[?10]$
$12.\mathbf{P}\text{-}R^{\exists}(v,r):(\forall x:A)C(x,v(x))\text{-}[11]$
$13.\mathbf{O}\text{-}R^{\exists}(v,r)/?\text{-}[?12]$
$14.\mathbf{P}\text{-}r:(\forall x:A)C(x,v(x))\text{-}[13]$
$15.\mathbf{O}\text{-}L^{\forall}(r):A\text{-}[?14]$
$16.\mathbf{P}\text{-}L^{\forall}(r)/?\text{-}[?15]$
$17.\mathbf{O}\text{-}w:A\text{-}[16]$
$18.\mathbf{P}\text{-}L^{\forall}(v):A\text{-}[?5]$
$19.\mathbf{O}\text{-}L^{\forall}(v)/?\text{-}[?18]$
$20.\mathbf{P}\text{-}w:A\text{-}[19]$
$21.\mathbf{O}\text{-}R^{\forall}(v):(\exists y:B(w))C(w,y)\text{-}[18]$
$22.\mathbf{P}\text{-}R^{\forall}(v)/?\text{-}[?21]$
$23.\mathbf{O}\text{-}(t_1,t_2):(\exists y:B(w))C(w,y)\text{-}[22]$
$24.\mathbf{P}\text{-}?_L\text{-}[?23]$
$25.\mathbf{O}\text{-}L^{\exists}(t_1,t_2):B(w)\text{-}[24]$
$26.\mathbf{P}\text{-}L^{\exists}(t_1,t_2)/?\text{-}[?25]$
$27.\mathbf{O}\text{-}t_1:B(w)\text{-}[26]$
$28.\mathbf{P}\text{-}?_R\text{-}[?23]$
$29.\mathbf{O}\text{-}R^{\exists}(t_1,t_2):C(w,t_1)\text{-}[28]$
$30.\mathbf{P}\text{-}R^{\exists}(t_1,t_2)/?\text{-}[?29]$
$31.\mathbf{O}\text{-}t_2:C(w,t_1)\text{-}[30]$
$32.\mathbf{P}\text{-}R^{\forall}(r):C(w,v(w))\text{-}[15]$
$33.\mathbf{O}\text{-}R^{\forall}(r)/?\text{-}[?32]$
$34.\mathbf{P}\text{-}t_2:C(w,v(w))\text{-}[33]$
$35.\mathbf{O}\text{-}v(w)/?\text{-}[?34]$
$36.\mathbf{P}\text{-}v(w)=t_1:B(w)\text{-}[?H1]$
$37.\mathbf{O}\text{-}v(w)=t_1:B(w)?\text{-}[?36]$
$38.\mathbf{P}\text{-}?_{\Pi-eq}\text{-}[?5,20\text{-}21,27]$
$39.\mathbf{O}\text{-}v(w)=t_1:B(w)\text{-}[38]$
$40.\mathbf{P}\text{-}sic\,(39)\text{-}[37]$
$41.\mathbf{O}\text{-}C(w,t_1)=C(w,t_1/v(w)):set\text{-}[36]$
$42.\mathbf{P}\text{-}t_2:C(w,t_1)<C(w,t_1)=C(w,t_1/v(w)):set>\text{-}[35]$

Fig. 4.1 Core of the winning **P**-strategy

4.2 From the Dialogical Strategy to the CTT Demonstration of the Axiom of Choice

1. (Initial stage) According to the algorithm developed in Chap. 3, we start by writing the thesis as the conclusion of the demonstration and the Opponent's initial concessions as its top assumptions. Recall that we do not keep the player-labels. We see from the core that the thesis is not case-dependent and is thus the result of an Introduction rule, here for material implication. This Introduction is displayed by **O**'s challenge 3, which constitutes the assumption, and **P**'s defensive move 6, which constitutes the premiss depending on this assumption. All together this yields Fig. 4.2.

$$C(x,y) : set\ (x:A, y:B(x)) \quad B(x) : set\ (x:A) \quad A : set$$
$$\vdots$$
$$L^{\rightarrow}(p)/v : (\forall x:A)(\exists y:B(x))C(x,y)^{(1)}$$
$$\vdots$$
$$\frac{R^{\rightarrow}(p)/(v,r) : (\exists f:(\forall x:A)B(x))(\forall x:A)C(x,f(x))}{p : (\forall x:A)(\exists y:B(x))C(x,y) \rightarrow (\exists f:(\forall x:A)B(x))(\forall x:A)C(x,f(x))} \rightarrow I(1)$$

Fig. 4.2 From the core to the demonstration (Initial stage)

Remark The substitutions $L^{\rightarrow}(p)/v$ and $R^{\rightarrow}(p)/(v, r)$ result from the moves 4-5 and 7-8. For the sake of simplification we adopt the following convention: once such a substitution has been established at, say, step n in some branch of the demonstration, we delete in the further steps (of that branch) the instruction and replace it with the play-object resulting from the substitution move.

2. Move 8 (in the core)—which resulted from the substitution $R^{\rightarrow}(p)/(v, r)$—is an existential posit by **P** that is not case-dependent. Thus, in the demonstration, it is caused by the application of an Introduction rule for existential quantification. Accordingly, the demonstration splits between the left-part and the right-part, similarly to what happens in the core at the point where two different branches are triggered by two different choices of **O** at move 9. This results in Fig. 4.3, on p. 62.

Remarks (i) To save some space we temporarily refer to the Opponent's initial concessions by using their labels H1, H2 and H3. (ii) The ramification is made in the scope of all of the assumptions which must then be placed at the top of the two parts of the demonstration. (iii) The substitutions $L^{\exists}(v, r)/v$ and $R^{\exists}(v, r)/r$ are established by moves 11-12 of the left branch of the core and by moves 13-14 of the right branch.

$$\dfrac{\dfrac{\begin{matrix}H1\,H2\,H3\\ \vdots\\ v:(\forall x:A)(\exists y:B(x))C(x,y)^{(1)}\\ \vdots\\ \mathbf{L^{\exists}(v,r)/v:(\forall x:A)B(x)}\end{matrix} \qquad \begin{matrix}H1\,H2\,H3\\ \vdots\\ v:(\forall x:A)(\exists y:B(x))C(x,y)^{(1)}\\ \vdots\\ \mathbf{R^{\exists}(v,r)/r:(\forall x:A)C(x,L^{\exists}(v,r)(x))}\end{matrix}}{(v,r):(\exists f:(\forall x:A)B(x))(\forall x:A)C(x,f(x))}\,\exists I}{p:(\forall x:A)(\exists y:B(x))C(x,y)\rightarrow(\exists f:(\forall x:A)B(x))(\forall x:A)C(x,f(x))}\,{\rightarrow}I(1)$$

Fig. 4.3 From the core to the demonstration (Stage 2)

3. Let us first develop the left part of the demonstration, though directly replacing the instructions $L^{\exists}(v,r)$ and $R^{\exists}(v,r)$ by the corresponding play-objects.

3.1. We now deal with $v : (\forall x : A)B(x)$ uttered by **P** at move 12. It results from the application of an Introduction rule: the **O**-challenge is move 13 and is introduced as a new assumption in this part of the demonstration; the Proponent's defence is move 26 and is introduced as the premiss depending on the new assumption.

The substitutions $L^{\forall}(v)/w$ and $R^{\forall}(v)/t_1$ are established respectively by **O** with moves 13-14-15, and by **P** with moves 26-27-28. All this together yields Fig. 4.4.

$$\dfrac{\dfrac{\dfrac{\begin{matrix}H1\,H2\,H3\\ \vdots\\ v:(\forall x:A)(\exists y:B(x))C(x,y)^{(1)}\\ \vdots\\ \mathbf{L^{\forall}(v)/w:A^{(2)}}\\ \vdots\\ \mathbf{R^{\forall}(v)/t_1:B(w)}\end{matrix}}{v:(\forall x:A)B(x)}\,\forall I(2) \qquad \begin{matrix}H1\,H2\,H3\\ \vdots\\ v:(\forall x:A)(\exists y:B(x))C(x,y)^{(1)}\\ \vdots\\ r:(\forall x:A)C(x,L^{\exists}(v,r)(x))\end{matrix}}{(v,r):(\exists f:(\forall x:A)B(x))(\forall x:A)C(x,f(x))}\,\exists I}{p:(\forall x:A)(\exists y:B(x))C(x,y)\rightarrow(\exists f:(\forall x:A)B(x))(\forall x:A)C(x,f(x))}\,{\rightarrow}I(1)$$

Fig. 4.4 From the core to the demonstration (Stage 3.1)

3.2. The Opponent does not challenge the Proponent's atomic posit 28 because she cannot: she has already made the same posit, therefore the structural rule *SR*3 (Modified Formal Rule) forbids her to challenge it. As specified by our algorithm, we thus indicate in the demonstration that the last introduced posit $t_1 : B(w)$ is justified by the move 25 and the application of rule *SR*3. Since **O** established the substitution $L^{\exists}(t_1, t_2)/t_1$ at moves 24-25, this results in Fig. 4.5 on next page.

$$
\dfrac{\dfrac{\dfrac{\dfrac{\begin{matrix} H1\ H2\ H3 \\ \vdots \\ v:(\forall x:A)(\exists y:B(x))C(x,y)^{(1)} \\ \vdots \\ w:A^{(2)} \\ \vdots \\ \mathbf{L^{\exists}(t_1,t_2)/t_1 : B(w)} \end{matrix}}{t_1 : B(w)}\,\mathbf{SR3}}{v:(\forall x:A)B(x)}\,\forall I(2) \qquad \begin{matrix} H1\ H2\ H3 \\ \vdots \\ v:(\forall x:A)(\exists y:B(x))C(x,y)^{(1)} \\ \vdots \\ r:(\forall x:A)C(x,L^{\exists}(v,r)(x)) \end{matrix}}{(v,r):(\exists f:(\forall x:A)B(x))(\forall x:A)C(x,f(x))}\,\exists I}{p:(\forall x:A)(\exists y:B(x))C(x,y) \to (\exists f:(\forall x:A)B(x))(\forall x:A)C(x,f(x))}\,\to I(1)
$$

Fig. 4.5 From the core to the demonstration (Stage 3.2)

3.3. Now, we see from the core and from the instruction $L^{\exists}(t_1, t_2)$ that moves 23 and 25 are **O** defending her existential expression of move 19, thus responding to **P**'s challenge 22. In the demonstration, it corresponds to the result of applying the ∃-Elimination rule. Moreover, the substitution $R^{\forall}(v)/(t_1, t_2)$ is established by the moves 19-20-21. All this yields Fig. 4.6.

$$
\dfrac{\dfrac{\dfrac{\dfrac{\dfrac{\begin{matrix} H1\ H2\ H3 \\ \vdots \\ v:(\forall x:A)(\exists y:B(x))C(x,y)^{(1)} \\ \vdots \\ w:A^{(2)} \\ \vdots \\ \mathbf{R^{\forall}(v)/(t_1,t_2) : (\exists y : B(w))C(w,y)} \end{matrix}}{t_1 : B(w)}\,\mathbf{\exists E}}{t_1 : B(w)}\,SR3}{v:(\forall x:A)B(x)}\,\forall I(2) \qquad \begin{matrix} H1\ H2\ H3 \\ \vdots \\ v:(\forall x:A)(\exists y:B(x))C(x,y)^{(1)} \\ \vdots \\ r:(\forall x:A)C(x,L^{\exists}(v,r)(x)) \end{matrix}}{(v,r):(\exists f:(\forall x:A)B(x))(\forall x:A)C(x,f(x))}\,\exists I}{p:(\forall x:A)(\exists y:B(x))C(x,y) \to (\exists f:(\forall x:A)B(x))(\forall x:A)C(x,f(x))}\,\to I(1)
$$

Fig. 4.6 From the core to the demonstration (Stage 3.3)

3.4. Move 19, as informed by the instruction $R^{\forall}(v)$, defends the universal brought forward by **O** at moves 3 and 5 (assumption 1 in the demonstration) and challenged by **P** at move 16. Thus move 19 corresponds in the demonstration to the application of a universal for which the premisses are provided by moves 5 and 16. The substitution $L^{\forall}(v)/w$ is established by **P** with moves 16-17-18. The result is given in Fig. 4.7.

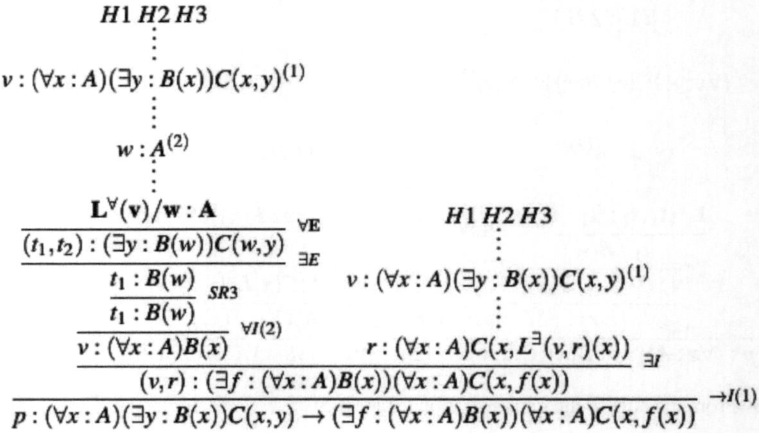

Fig. 4.7 From the core to the demonstration (Stage 3.4)

3.5. Similarly to what happened in step 3.2, the Opponent cannot challenge **P**'s move 18 because she herself made the same posit at move 15. In other words, **P**'s move 18 is justified by the structural rule *SR*3. Moreover, we have already included the Opponent's move 15 as the assumption 2 in the demonstration. See Fig. 4.8 below.

3.6. No other move from the core is to be included in this left part of the demonstration: we have dealt with all the moves resulting from the Opponent choosing at move 9 to ask for the left side of move 8. We can thus remove the vertical dots and link the assumptions 1 and H1-H3 to this part of the demonstration. When doing so we ensure that assumption 1 is correctly placed as a premiss of the ∀-Elimination rule triggered by it. We also delete the redundant **P** atomic expressions, namely those triggered by the *SR*3 rule. The left part of the demonstration is thus as in Fig. 4.9.

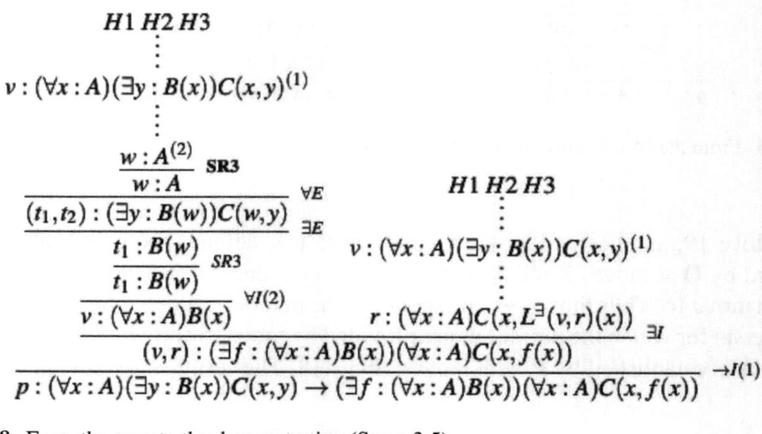

Fig. 4.8 From the core to the demonstration (Stage 3.5)

$$\dfrac{\dfrac{\dfrac{\dfrac{\dfrac{H1\ H2\ H3 \quad v:(\forall x:A)(\exists y:B(x))C(x,y)^{(1)} \quad w:A^{(2)}}{(t_1,t_2):(\exists y:B(w))C(w,y)}\,\forall E}{t_1:B(w)}\,\exists E}{v:(\forall x:A)B(x)}\,\forall I(2) \qquad \begin{array}{c} H1\ H2\ H3 \\ \vdots \\ v:(\forall x:A)(\exists y:B(x))C(x,y)^{(1)} \\ \vdots \\ r:(\forall x:A)C(x,L^{\exists}(v,r)(x)) \end{array}}{(v,r):(\exists f:(\forall x:A)B(x))(\forall x:A)C(x,f(x))}\,\exists I}{p:(\forall x:A)(\exists y:B(x))C(x,y)\rightarrow(\exists f:(\forall x:A)B(x))(\forall x:A)C(x,f(x))}\,\rightarrow I(1)$$

Fig. 4.9 From the core to the demonstration (Stage 3.6)

4. We now develop the right-hand part of the demonstration. Because of the limited space, we shall temporarily refer to the left-hand part by using [*Left*]. We will bring everything together later on.

4.1. We pick up the right-hand part at move 10 which brings a universal expression forward in the second branch of the core. We first implement the substitution $L^{\exists}(v,r)/v$ at the right of the colon, as established by moves 10-11-12, which gets us to the Proponent's move 14. Since it is not a case-dependent **P**-move, it proceeds from the application of a ∀-Introduction rule in the demonstration. The Opponent's move 15 (assumption 3) provides the assumption for the application of this rule and the premiss depending on assumption 3 is provided by **P**'s defensive move 32. Furthermore, moves 15-16-17 and 32-33-34 supply the substitutions $L^{\forall}(r)/w$ and $R^{\forall}(r)/t_2$, respectively provided by **O** and **P**. The right-hand part of the demonstration is thus developed as in Fig. 4.10 below.

$$\dfrac{\dfrac{\begin{array}{c} [Left] \\ \vdots \end{array} \qquad \dfrac{\begin{array}{c} H1\ H2\ H3 \\ \vdots \\ v:(\forall x:A)(\exists y:B(x))C(x,y)^{(1)} \\ \vdots \\ \mathbf{L^{\forall}(v)/w:A^{(3)}} \\ \vdots \\ \mathbf{R^{\forall}(r)/t_2:C(w,v(w))} \end{array}}{r:(\forall x:A)C(x,\mathbf{L^{\exists}(v,r)/v}(x))}\,\forall\mathbf{I(3)}}{(v,r):(\exists f:(\forall x:A)B(x))(\forall x:A)C(x,f(x))}\,\exists I}{p:(\forall x:A)(\exists y:B(x))C(x,y)\rightarrow(\exists f:(\forall x:A)B(x))(\forall x:A)C(x,f(x))}\,\rightarrow I(1)$$

Fig. 4.10 From the core to the demonstration (Stage 4.1)

4.2. Move 34 of the core is challenged by the Opponent who asks to substitute $v(w)$ with a suitable play-object—see in Sect. 2.3 the substitution rule for functions. The Proponent's defence at move 42 shows that his move 34 was justified by the equality $C(w, t_1) = C(w, t_1/v(w)) : set$ and by the posit $t_2 : C(w, t_1)$. In the demonstration, it corresponds to an application of the CTT rule of Set Equality. Since this rule requires two premisses, it triggers a ramification in the demonstration and we thus have to place the current assumptions in both parts. This yields the result shown in Fig. 4.11, also below.

$$H1\; H2\; H3 \qquad\qquad H1\; H2\; H3$$

$$v : (\forall x : A)(\exists y : B(x))C(x,y)^{(1)} \qquad v : (\forall x : A)(\exists y : B(x))C(x,y)^{(1)}$$

$$w : A^{(3)} \qquad\qquad w : A^{(3)}$$

$$[Left] \quad \cfrac{\cfrac{\cfrac{\cfrac{\mathbf{C(w,t_1) = C(w,t_1/v(w)) : set} \qquad \mathbf{t_2 : C(w,t_1)}}{t_2 : C(w,v(w))}\,\text{Set-Eq}}{r : (\forall x : A)C(x,v(x))}\,\forall I(3)}{(v,r) : (\exists f : (\forall x : A)B(x))(\forall x : A)C(x,f(x))}\,\exists I}{p : (\forall x : A)(\exists y : B(x))C(x,y) \rightarrow (\exists f : (\forall x : A)B(x))(\forall x : A)C(x,f(x))}\,\rightarrow I(1)$$

Fig. 4.11 From the core to the demonstration (Stage 4.2)

Remark and explanation on Fig. 4.11 (*Stage 4.2*): There is no ramification in the core reflecting this ramification in the demonstration. However, the dialogical play *does* account for the twofold aspect of this application of the rule Set-Eq, since there are two aspects in the Proponent's defence 42, and both are dealt with in the play:

(a) The equality $C(w, t_1) = C(w, t_1/v(w)) : set$, which **P** must establish by making **O** posit it herself first: this is the purpose of moves 36 to 41, and we treat them in the next steps numbered 4.3.X.
(b) The atomic expression $t_2 : C(w, t_1)$. This simply involves the use of the Modified Formal Rule *SR*3 in relation to **O**'s move 31, and we will come back to it afterwards in steps 4.4.X.

4.3. We now focus on the left-hand part of the new ramification, that is to say to the middle branch of the demonstration. This requires some special attention so we will describe it progressively. Moreover, we will use [*Right*] for the far right branch.

4.3.1. As we explained after the second play in Sect. 4.1.1, the reason why **O** cannot question the proviso $< C(w, t_1) = C(w, t_1/v(w)) : set >$ is that she posited it herself at move 41. Thus we are dealing with an application of *SR*3 as displayed in Fig. 4.12.

$$
\begin{array}{c}
H1\;H2\;H3 \\
\vdots \\
v:(\forall x:A)(\exists y:B(x))C(x,y)^{(1)} \\
\vdots \\
w:A^{(3)} \\
\vdots
\end{array}
$$

$$
\dfrac{\dfrac{[Left] \quad \vdots \qquad \dfrac{\dfrac{\dfrac{\mathbf{C(w,t_1)=C(w,t_1/v(w)):set}}{C(w,t_1)=C(w,t_1/v(w)):set}\,\mathbf{SR3} \qquad \begin{array}{c}[Right]\\ \vdots\end{array}}{t_2:C(w,v(w))}\,\textit{Set-Eq}}{r:(\forall x:A)C(x,v(x))}\,\forall I(3)}{(v,r):(\exists f:(\forall x:A)B(x))(\forall x:A)C(x,f(x))}\,\exists I}{p:(\forall x:A)(\exists y:B(x))C(x,y)\to(\exists f:(\forall x:A)B(x))(\forall x:A)C(x,f(x))}\,\to I(1)
$$

Fig. 4.12 From the core to the demonstration (Stage 4.3.1)

4.3.2. The Opponent's move 41 is a defence in response to **P**'s challenge 36. The challenge consists in asking for a substitution in $C(x, y) : set(x : A, y : B(x))$, namely $H1$, by positing the equality $v(w) = t_1 : B(w)$.

Again, we temporarily write $H1$ for $C(x, y) : set\ (x : A, y : B(x))$. Since it is the only assumption relevant for the application of the Substitution rule, we do not rewrite the other ones in this new part of the demonstration. All this yields Fig. 4.13.

$$
\begin{array}{c}
H1\;H2\;H3 \\
\vdots \\
v:(\forall x:A)(\exists y:B(x))C(x,y)^{(1)} \\
\vdots \\
w:A^{(3)} \\
\vdots
\end{array}
$$

$$
\dfrac{\dfrac{[Left] \quad \vdots \qquad \dfrac{\dfrac{\dfrac{\dfrac{\mathbf{v(w)=t_1:B(w)} \qquad \mathbf{H1}}{C(w,t_1)=C(w,t_1/v(w)):set}\,\mathbf{Subst}}{C(w,t_1)=C(w,t_1/v(w)):set}\,SR3 \qquad \begin{array}{c}[Right]\\ \vdots\end{array}}{t_2:C(w,v(w))}\,\textit{Set-Eq}}{r:(\forall x:A)C(x,v(x))}\,\forall I(3)}{(v,r):(\exists f:(\forall x:A)B(x))(\forall x:A)C(x,f(x))}\,\exists I}{p:(\forall x:A)(\exists y:B(x))C(x,y)\to(\exists f:(\forall x:A)B(x))(\forall x:A)C(x,f(x))}\,\to I(1)
$$

Fig. 4.13 From the core to the demonstration (Stage 4.3.2)

4.3.3. But before defending, **O** counterattacks move 36 by asking for a justification of the equality posited by **P**. The Proponent then makes her posit it herself so that he can defend at move 40 with "*sic* (39)". As for the way **P** does that, we leave it for the next step. For now, the point is that move 36 is thus justified by the structural rule *SR*3 and the demonstration develops as shown in Fig. 4.14.

$$H1\ H2\ H3$$
$$\vdots$$
$$v : (\forall x : A)(\exists y : B(x))C(x,y)^{(1)}$$
$$\vdots$$
$$w : A^{(3)}$$
$$\vdots$$
$$\dfrac{\dfrac{\dfrac{\dfrac{\mathbf{v(w) = t_1 : B(w)}}{v(w) = t_1 : B(w)}\,\mathit{SR3} \qquad H1}{C(w,t_1) = C(w,t_1/v(w)) : set}\,\mathit{Subst}}{C(w,t_1) = C(w,t_1/v(w)) : set}\,\mathit{SR3} \qquad \begin{matrix}[Right]\\ \vdots\end{matrix}$$
$$\dfrac{\dfrac{\dfrac{\begin{matrix}[Left]\\ \vdots\end{matrix} \qquad \dfrac{\dots}{t_2 : C(w,v(w))}\,\mathit{Set\text{-}Eq}}{r : (\forall x : A)C(x,v(x))}\,\forall I(3)}{(v,r) : (\exists f : (\forall x : A)B(x))(\forall x : A)C(x,f(x))}\,\exists I}{p : (\forall x : A)(\exists y : B(x))C(x,y) \to (\exists f : (\forall x : A)B(x))(\forall x : A)C(x,f(x))}\,\to I(1)$$

Fig. 4.14 From the core to the demonstration (Stage 4.3.3)

4.3.4. Let us now look at the way **P** makes **O** posit $v(w) = t_1 : B(w)$ at move 39. He applies the dialogical rule for Π-Equality, taking advantage of the Opponent's moves 21 and 27 (under the assumption 1 given at move 5) as well as his own posit $w : A$ at move 20. A crucial point is that the Π-Equality rule is applied in relation to the function v which **P** has chosen with the moves 10-11-12. To apply the corresponding CTT rule, it is necessary to first introduce the function explicitly: we will come back to the way the Opponent's moves 21 and 27 are related to this Introduction in the next steps. Thus, the demonstration is developed as in Fig. 4.15 below.

$$H1\ H2\ H3$$
$$\vdots$$
$$v : (\forall x : A)(\exists y : B(x))C(x,y)^{(1)}$$
$$\vdots$$
$$w : A^{(3)}$$
$$\vdots$$
$$\dfrac{\dfrac{\dfrac{\dfrac{\dfrac{\mathbf{v : (\forall x : A)B(x)} \qquad \mathbf{w : A}}{v(w) = t_1 : B(w)}\,\Pi Eq}{v(w) = t_1 : B(w)}\,\mathit{SR3} \qquad H1}{C(w,t_1) = C(w,t_1/v(w)) : set}\,\mathit{Subst}}{C(w,t_1) = C(w,t_1/v(w)) : set}\,\mathit{SR3} \qquad \begin{matrix}[Right]\\ \vdots\end{matrix}$$
$$\dfrac{\dfrac{\dfrac{\begin{matrix}[Left]\\ \vdots\end{matrix} \qquad \dfrac{\dots}{t_2 : C(w,v(w))}\,\mathit{Set\text{-}Eq}}{r : (\forall x : A)C(x,v(x))}\,\forall I(3)}{(v,r) : (\exists f : (\forall x : A)B(x))(\forall x : A)C(x,f(x))}\,\exists I}{p : (\forall x : A)(\exists y : B(x))C(x,y) \to (\exists f : (\forall x : A)B(x))(\forall x : A)C(x,f(x))}\,\to I(1)$$

Fig. 4.15 From the core to the demonstration (Stage 4.3.4)

4.3.5. Let us also deal with **P**'s move 20: he is entitled to play it because **O** made the same posit at move 17 (assumption 3), therefore it results from an application of *SR*3.

Now, let us consider the introduction of the function $v : (\forall x : A)B(x)$. The reason why **P** chose t_1 as a play-object for $v(w)$ is that the Opponent herself posited $t_1 : B(w)$ at move 27. Also, and more important, this posit is made under the assumption that $w : A$, which the Opponent conceded herself (assumption 3 in the demonstration). Thus the choice of the function v by **P** can be said to reflect an application of the ∀-Introduction rule from these two premisses.

Given that the substitution $L^{\exists}(t_1, t_2)/t_1$ is established with moves 25-26-27, all this yields Fig. 4.16.

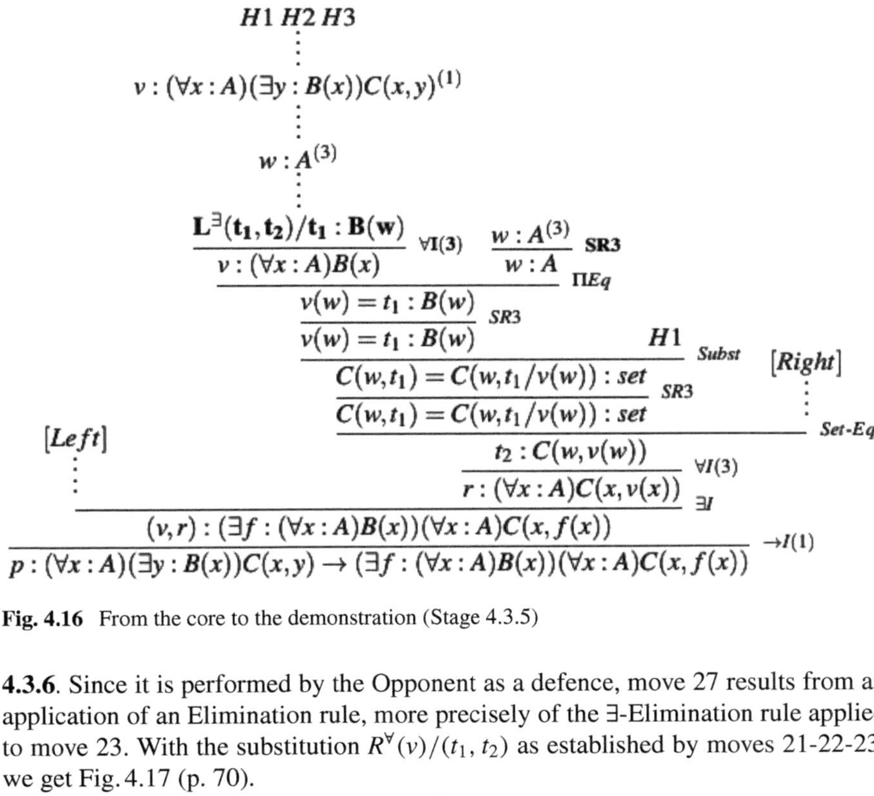

Fig. 4.16 From the core to the demonstration (Stage 4.3.5)

4.3.6. Since it is performed by the Opponent as a defence, move 27 results from an application of an Elimination rule, more precisely of the ∃-Elimination rule applied to move 23. With the substitution $R^{\forall}(v)/(t_1, t_2)$ as established by moves 21-22-23, we get Fig. 4.17 (p. 70).

4.3.7. Similarly, the expression added in the previous stage results from an application of the ∀-Elimination rule: the challenged **O**-move is assumption 1 (move 5) and the Proponent's challenge is move 18. The substitution $L^{\forall}(v)/w$ is established by **P** at moves 18-19-20. We have also seen previously that the Proponent's move 20 is allowed by *SR*3 because **O** made the same posit at move 17.

From all this we get Fig. 4.18 (also on p. 70).

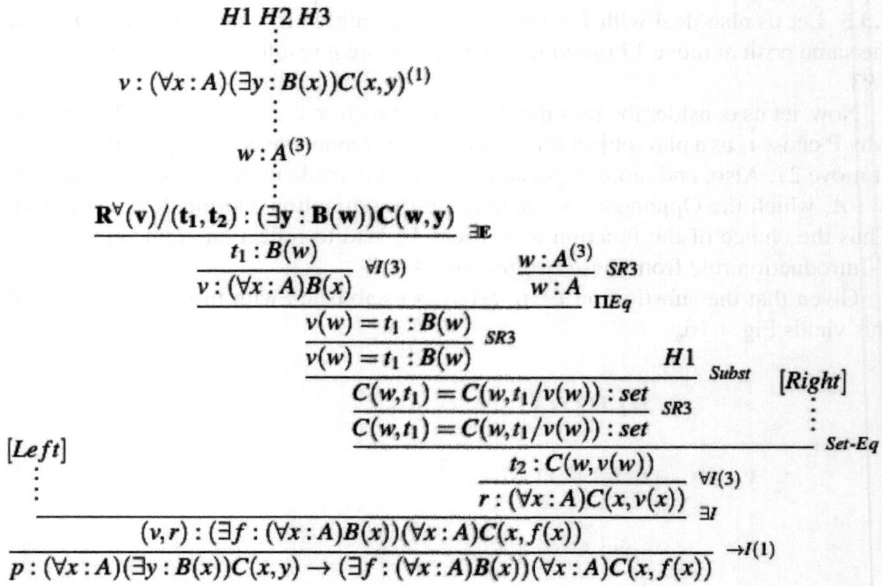

Fig. 4.17 From the core to the demonstration (Stage 4.3.6)

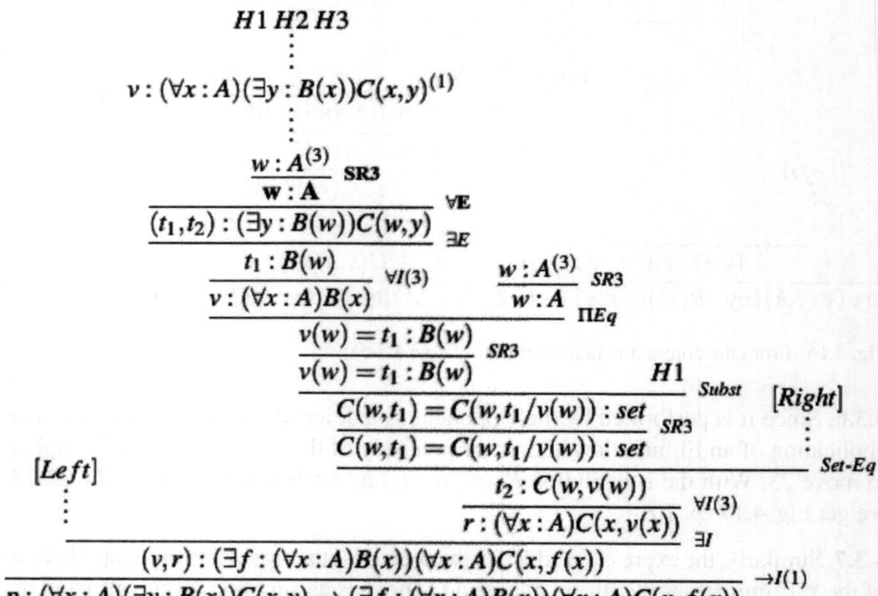

Fig. 4.18 From the core to the demonstration (Stage 4.3.7)

4.3.8. There is no other move to be included from the core in this part of the demonstration: we have dealt with the whole chain of moves involved in the justification of the proviso $< C(w, t_1) = C(w, t_1/v(w)) : set >$ in the Proponent's move 42. We can remove the remaining vertical dots, and the applications of rule *SR*3 as well as the redundant expressions they produced.

This yields Fig. 4.19, ending the process initiated in stage 4.3.1 about what we then called the central branch of the demonstration.

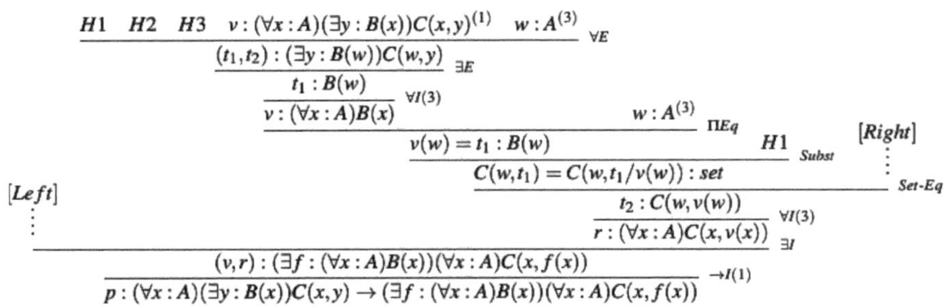

Fig. 4.19 From the core to the demonstration (Stage 4.3.8)

4.4. Let us develop now the last part of the demonstration: its far right. To save space we temporarily refer to the part we have treated in steps 4.3.X as [*Subst*].

4.4.1. We left this far right-hand part with the other aspect of move 42, namely the posit of the atomic expression $t_2 : C(w, t_1)$. Similarly to what has now happened several times, **O** cannot challenge the posit because she posited it herself at move 31—after the substitution $R^{\exists}(t_1, t_2)/t_2$ is established with moves 29-30-31. The far right-hand part of the demonstration thus becomes as in Fig. 4.20.

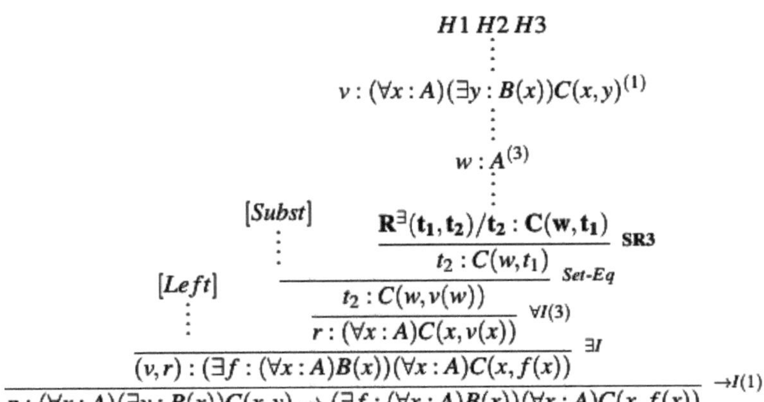

Fig. 4.20 From the core to the demonstration (Stage 4.4.1)

4.4.2. This in turn is a defensive move by **O**, thus resulting from an application of the ∃-Elimination rule on her move 23, after moves 21-22-23 established the substitution $R^{\forall}(v)/(t_1, t_2)$. The result is shown in Fig. 4.21.

$$H1\ H2\ H3$$
$$\vdots$$
$$v : (\forall x : A)(\exists y : B(x))C(x,y)^{(1)}$$
$$\vdots$$
$$w : A^{(3)}$$
$$\vdots$$
$$\mathbf{R^{\forall}(v)/(t_1,t_2) : (\exists y : B(w))C(w,y)}$$
$$\frac{[Subst] \quad \vdots \quad \frac{\frac{\mathbf{R^{\forall}(v)/(t_1,t_2) : (\exists y : B(w))C(w,y)}}{t_2 : C(w,t_1)}\ \exists E}{t_2 : C(w,t_1)}\ SR3}{t_2 : C(w, v(w))}\ \textit{Set-Eq}$$
$$\frac{t_2 : C(w, v(w))}{r : (\forall x : A)C(x, v(x))}\ \forall I(3)$$
$$\frac{[Left] \quad \vdots \quad r : (\forall x : A)C(x, v(x))}{(v,r) : (\exists f : (\forall x : A)B(x))(\forall x : A)C(x, f(x))}\ \exists I$$
$$\frac{(v,r) : (\exists f : (\forall x : A)B(x))(\forall x : A)C(x, f(x))}{p : (\forall x : A)(\exists y : B(x))C(x,y) \rightarrow (\exists f : (\forall x : A)B(x))(\forall x : A)C(x, f(x))}\ {\rightarrow}I(1)$$

Fig. 4.21 From the core to the demonstration (Stage 4.4.2)

4.4.3. As informed by the instruction in the substitution, move 23 is a defence after a **P**-challenge on the universal which was brought forward by **O** at move 5 (assumption 1). The Proponent's challenge is made at move 18 which, after the resolution of the instruction, yields move 20 ($w : A$). Once again this **P**-challenge is itself justified by the structural rule $SR3$ since **O** made the same posit at move 17 (assumption 3). See Fig. 4.22 below.

$$H1\ H2\ H3$$
$$\vdots$$
$$v : (\forall x : A)(\exists y : B(x))C(x,y)^{(1)}$$
$$\vdots$$
$$\frac{w : A^{(3)}}{\mathbf{L^{\forall}(v)/w : A}}\ \mathbf{SR3}$$
$$\frac{\mathbf{L^{\forall}(v)/w : A}}{(t_1,t_2) : (\exists y : B(w))C(w,y)}\ \forall E$$
$$\frac{[Subst] \quad \vdots \quad \frac{\frac{(t_1,t_2) : (\exists y : B(w))C(w,y)}{t_2 : C(w,t_1)}\ \exists E}{t_2 : C(w,t_1)}\ SR3}{t_2 : C(w, v(w))}\ \textit{Set-Eq}$$
$$\frac{t_2 : C(w, v(w))}{r : (\forall x : A)C(x, v(x))}\ \forall I(3)$$
$$\frac{[Left] \quad \vdots \quad r : (\forall x : A)C(x, v(x))}{(v,r) : (\exists f : (\forall x : A)B(x))(\forall x : A)C(x, f(x))}\ \exists I$$
$$\frac{(v,r) : (\exists f : (\forall x : A)B(x))(\forall x : A)C(x, f(x))}{p : (\forall x : A)(\exists y : B(x))C(x,y) \rightarrow (\exists f : (\forall x : A)B(x))(\forall x : A)C(x, f(x))}\ {\rightarrow}I(1)$$

Fig. 4.22 From the core to the demonstration (Stage 4.4.3)

4.4.4. We have dealt with the whole chain of moves in the core which are involved in the justification of the Proponent's posit $t_2 : C(w, t_1)$ at move 42. There are thus no other moves to include in this part of the demonstration. Figure 4.23 shows the right-hand part of the demonstration as it results after removing the vertical dots, as well as applications of *SR*3 and redundant expressions they produced.

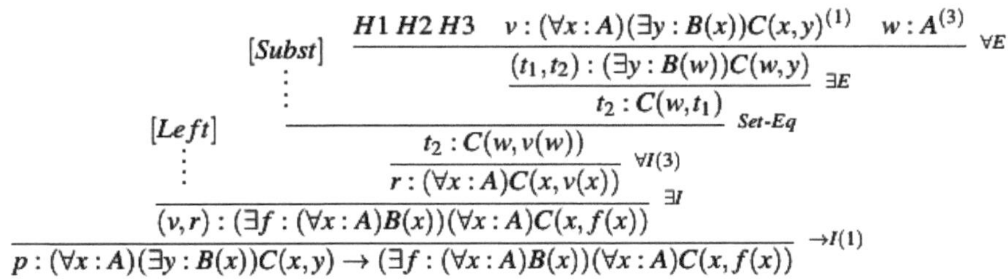

Fig. 4.23 From the core to the demonstration (Stage 4.4.4)

We have now reached the end of the construction of the demonstration structure, and it is time to write it down without using the abreviations [*Left*] and [*Subst*]. The resulting structure is given in Fig. 4.24, p. 74. However, the result is not yet written in the language of CTT. To achieve this two rewriting steps are required. In the first one, we re-introduce instructions and in the second step we translate these instructions into proof-objects. In principle, we could have chosen to define our procedure in such a way that we only keep the instructions from the start and save a step. The problem is that the notation becomes really heavy and quite difficult to follow. Let us tackle now the task of reintroducing the instructions.

5. Looking through the history of substitutions, such as $L^{\rightarrow}(p)/v$, that we established during the construction of the demonstration, we recall that:

i. $v/L^{\rightarrow}(p)$, and $(v, r)/R^{\rightarrow}(p)$
ii. $v/L^{\exists}(v, r)$ and thus $v/L^{\exists}(R^{\rightarrow}(p))$
   $r/R^{\exists}(v, r)$ and thus $r/R^{\exists}(R^{\rightarrow}(p))$
iii. $w/L^{\forall}(v)$ and thus $w/L^{\forall}(L^{\exists}(R^{\rightarrow}(p)))$
   $(t_1, t_2)/R^{\forall}(v)$ and thus $(t_1, t_2)/R^{\forall}(L^{\rightarrow}(p))$
iv. $t_1/L^{\exists}(t_1, t_2)$ and thus $t_1/L^{\exists}(R^{\forall}(L^{\rightarrow}(p)))$
   $t_2/R^{\exists}(t_1, t_2)$ and thus $t_2/R^{\exists}(R^{\forall}(L^{\rightarrow}(p)))$
v. An important point is that the expressions we removed as redundant, i.e., the atomic expressions posited by **P** according to the rule *SR*3, established other substitutions in the cases of t_1, t_2 and w:
   $t_1/R^{\forall}(v)$ and thus $t_1/R^{\forall}(L^{\exists}(R^{\rightarrow}(p)))$,
   $t_2/R^{\forall}(r)$ and thus $t_2/R^{\forall}(R^{\exists}(R^{\rightarrow}(p)))$,
   $w/L^{\forall}(v)$ and thus $w/L^{\forall}(L^{\rightarrow}(p))$.

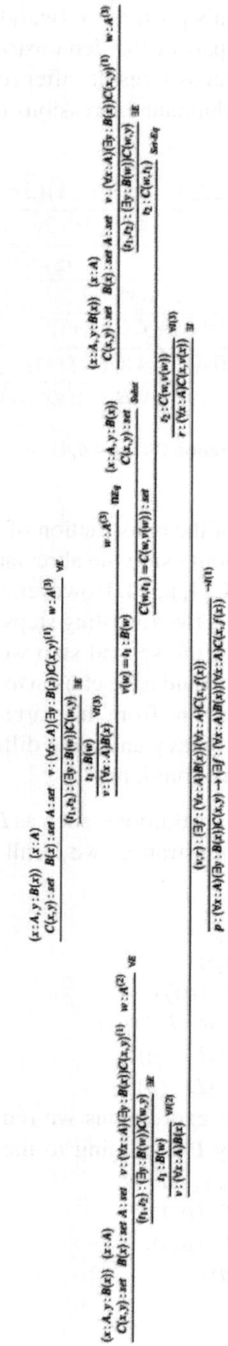

Fig. 4.24 From the core to the demonstration: after going through stages 1 to 4.4.4

Recall that no instruction has been introduced for the play-object p occurring in the thesis. The substitutions in (iv) and (v) illustrate how heavy the procedure would have been, had we kept instructions instead of play-objects.

In the final stage of the procedure, particular attention must be given to the substitutions described in (v). Indeed, in the course of a demonstration, expressions in which the play-objects t_1, t_2 and w occur might be used several times while applying various rules. Consider for example the assumption $w : A$ which is sometimes used to apply $\forall$-Elimination on assumption (1) and sometimes used to apply $\forall$-Introduction—for example to conclude $v : (\forall x : A)B(x)$. This duality is in fact the very reason why some expressions might occur twice in a dialogical play as posited once by **O** and another time by **P** (a distinction of no use in CTT). It is also the manifestation of the basic principle behind **P**'s winning strategy, namely choosing smartly the same play-objects for his own instructions that **O** chose for hers.

Given the size of Fig. 4.24, it is not possible to write entirely out the result of substituting play-objects with the instructions according to the list i-v above. For this reason, we give in Fig. 4.25 the result only for the far left of the demonstration—the part we have called [*Left*] during the translation procedure.

$$\cfrac{\begin{array}{cccc} \begin{array}{c}(x:A,\, y:B(x))\\ C(x,y):set\end{array} & \begin{array}{c}(x:A)\\ B(x):set\end{array} \; A:set & L^{\rightarrow}(p):(\forall x:A)(\exists y:B(x))C(x,y)^{(1)} & L^{\forall}(L^{\rightarrow}(p))[L^{\forall}(L^{\exists}(R^{\rightarrow}(p)))]:A^{(2)}\end{array}}{R^{\forall}(L^{\rightarrow}(p)):(\exists y:B(L^{\forall}(L^{\rightarrow}(p))))C(L^{\forall}(L^{\rightarrow}(p)),y)}\;\forall E$$

$$\cfrac{R^{\forall}(L^{\rightarrow}(p)):(\exists y:B(L^{\forall}(L^{\rightarrow}(p))))C(L^{\forall}(L^{\rightarrow}(p)),y)}{L^{\exists}(R^{\forall}(L^{\rightarrow}(p)))[R^{\forall}(L^{\exists}(R^{\rightarrow}(p)))]:B(L^{\forall}(L^{\rightarrow}(p)))}\;\exists E$$

$$\cfrac{L^{\exists}(R^{\forall}(L^{\rightarrow}(p)))[R^{\forall}(L^{\exists}(R^{\rightarrow}(p)))]:B(L^{\forall}(L^{\rightarrow}(p)))}{L^{\exists}(R^{\rightarrow}(p)):(\forall x:A)B(x)}\;\forall I(2)$$

$$\cfrac{L^{\exists}(R^{\rightarrow}(p)):(\forall x:A)B(x) \qquad \vdots}{R^{\rightarrow}(p):(\exists f:(\forall x:A)B(x))C(x,f(x))}\;\exists I$$

$$\cfrac{R^{\rightarrow}(p):(\exists f:(\forall x:A)B(x))C(x,f(x))}{p:(\forall x:A)(\exists y:B(x))C(x,y)\rightarrow(\exists f:(\forall x:A)B(x))(\forall x:A)C(x,f(x))}\;\rightarrow I(1)$$

Fig. 4.25 From the core to the demonstration: after Stage 5 (partial view)

6. We are one step away from getting to the proper CTT demonstration of the Axiom of Choice: the only thing that remains is to go from dialogical instructions to CTT proof-objects. However, it is quite easy since we know which rules have been applied, and also because the syntax of nested instructions allows to keep track of the objects on which these rules operate.

To replace instructions with proof-objects, we start with the assumptions and go down in the demonstration. Thus, we first interpret the instruction $L^{\rightarrow}(p)$ as the proof-object z and $L^{\forall}(L^{\rightarrow}(p))$ as x. We chose these particular names because they are the ones chosen by Martin-Löf in his own demonstration. Anyway, the important thing is that z and x are arbitrary: we assume an arbitrary x in A and an arbitrary method z to get objects of type $(\exists y : B(x))C(x, y)$, given objects x in A. From this we get

1. $R^{\forall}(L^{\rightarrow}(p))$ is the result of applying $L^{\rightarrow}(p)$ (namely z) to $L^{\forall}(L^{\rightarrow}(p))$ (namely x). That is, $R^{\forall}(L^{\rightarrow}(p))$ is understood as $Ap(z, x)$.
2. Next we have an application of the $\exists$-Elimination rule, so the dialogical $L^{\exists}$ is to be understood as left-projection. Hence $L^{\exists}(R^{\forall}(L^{\rightarrow}(p)))$ becomes $p(Ap(z, x))$.

3. After that we have an application of ∀-Introduction, using assumption (2). For this reason we must understand $L^{\exists}(R^{\rightarrow}(p))$ as a lambda-abstraction on x, i.e., as $(\lambda x)p(Ap(z,x))$.

Explanation: It is probably a little confusing that an occurrence of $L^{\exists}$ was interpreted as left projection in 2 whereas another occurrence is interpreted as lambda-abstraction in 3. The reason for this is the following. The dialogical instruction $L^{\exists}(R^{\rightarrow}(p))$ comes from the fact that the corresponding move was a **P**-defence of an existential, and it actually accounts for the fact that the proof-object we will associate to it will serve as the first member (the left-projection) of a pair in the next step of the demonstration. But we are now interested in the demonstration alone and reading it from top to bottom. So this last remark is an anticipation on what will happen afterwards. At this point of the procedure, we are only interested in the fact that the rule applied here is a ∀-Introduction rule. This is why the dialogical instruction must be understood as a lambda-abstraction.

We have reached the point where we need information from the two other branches of the demonstration before dealing with the next rule application.

4. We now consider the middle branch. It starts in the same way as the left branch we just treated, until the fifth member which results from application of the Π-Equality rule. At this step we have the application of $L^{\exists}(R^{\rightarrow}(p))$ to x equaling, in $B(x)$, to $L^{\exists}(R^{\forall}(L^{\rightarrow}(p)))$. Since we know the proof-objects associated with them, we can write the result of the substitution, which is: $Ap\big((\lambda x)p(Ap(z,x)),x\big) = p(Ap(z,x)) : B(x)$.
5. There are no new instructions to deal with in the next step of this branch, which results from applying the Substitution rule: the instructions occurring can be directly replaced with the proof-objects we have already associated them to. The result is thus $C(x, p(Ap(z,x))) = C(x, Ap((\lambda x)p(Ap(z,x)),x)) : set$.
6. Next is the right branch. It starts in the same way as the other branches, until its third member. Since we face an application of the ∃-Elimination rule, $R^{\exists}(R^{\forall}(L^{\rightarrow}(p)))$ is to be directly understood as the right-projection of $R^{\forall}(L^{\rightarrow}(p))$, which we have already replaced with $Ap(z,x)$. Thus we replace $R^{\exists}(R^{\forall}(L^{\rightarrow}(p)))$ with $q(Ap(z,x))$.

At this point we have reached the step where the middle and right branches join:

7. This is the step resulting from applying the Set-Equality rule. All the dialogical instructions occurring at this step have already been dealt with, so we simply replace them accordingly and get $q(Ap(z,x)) : C(x, Ap((\lambda x)p(Ap(z,x)),x))$.
8. The next step is the result of applying the ∀-Introduction rule. Thus, the dialogical instruction $R^{\exists}(R^{\rightarrow}(p))$ at the left of the colon is to be associated with a lambda-abstraction.
   This is the same phenomenon faced in 3: the dialogical instruction comes from a **P**-defence of an existential posit, but we see from the "top-down" reading of the demonstration that the CTT rule applied here is ∀-Introduction. The dialogical instruction anticipates on the next step of the demonstration and tells us that the proof-object will constitute the second member (right projection) of a pair.

Next the left and right parts join in the main body of the demonstration:

9. We know from the remarks we made in 3 and 8 that $R^{\rightarrow}(p)$ is the pair

$$\left(L^{\exists}(R^{\rightarrow}(p)), R^{\exists}(R^{\rightarrow}(p))\right)$$

Replacing these with the proof-objects we associated to them in 3 and 8, we see that $R^{\rightarrow}(p)$ must be associated with $\left((\lambda x)p(Ap(z,x)), (\lambda x)q(Ap(z,x))\right)$.

10. Finally, the conclusion of the demonstration results from a lambda-abstraction ($\rightarrow$ Introduction rule) with assumption (1). Thus, we can conclude that the proof-object which p stands for in the conclusion is:

$$(\lambda z)\left((\lambda x)p(Ap(z,x)), (\lambda x)q(Ap(z,x))\right)$$

The result of this procedure by which we replace dialogical instructions with CTT proof-objects is the demonstration of Fig. 4.26, p. 78, exactly as it is described in Martin-Löf (1984, pp. 50–51).

4.3 The Extensional Version of the Axiom of Choice

Why do constructivists reject the Axiom of Choice? Martin-Löf (2006) shows that what the constructivists reject is its extensional formulation. That is:

$$(\forall x : A)(\exists y : B(x))C(x,y) \rightarrow (\exists f : (\forall x : A)B(x))(Ext(f) \wedge (\forall x : A)C(x, f(x)))$$

where $Ext(f)$ is shorthand for $(\forall i, j : A)(i =_A j \rightarrow f(i) = f(j))$.

The point of the intensional version of the axiom, as explained in the previous sections, is that the intensionality allows to assert the existence of a choice function having a given property and which should be extracted from the antecedent. Certainly we cannot, from the fact that we can extract an intensional function from the antecedent of the implication, jump to conclude the function's extensionality.[7] In fact, as shown by Martin-Löf (2006), the extensional version even holds constructively *if*, in the antecedent, we assume that there exists a *unique* y for each x. That is, the following extensional formulation of the Axiom of Choice (assuming the uniqueness $\exists! y$) holds in Constructive Type Theory:

$$(\forall x : A)(\exists! y : B(x))C(x,y) \rightarrow (\exists f : (\forall x : A)B(x))(Ext(f) \wedge (\forall x : A)C(x, f(x)))$$

[7]Cf. Martin-Löf (2006, p. 349) with the following passage of Bernays (1930) quoted by Bell (2009, p. 207): *The existence of a function with a* [given] *property in no way guarantees the existence of concept formation through which a determinate function with* [that] *property is uniquely fixed.*

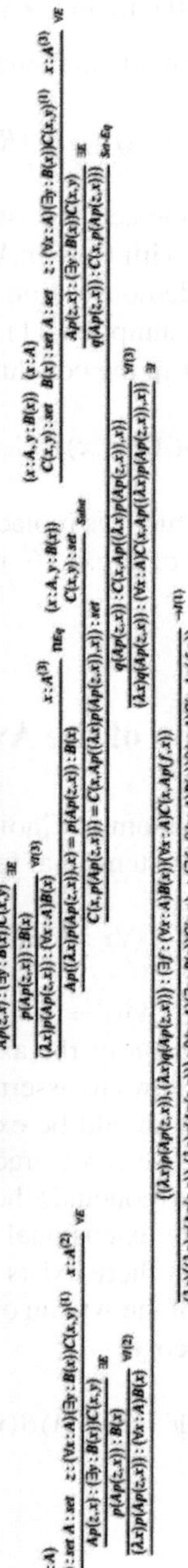

Fig. 4.26 The CTT demonstration obtained from Fig. 4.1 after stages 1 to 6 of Sect. 4.2

4.3.1 Martin-Löf on the Extensional Version with Uniqueness

The proof of Martin-Löf (2006, pp. 348–349) is straightforward and transparent. Let us spell the details out:

The assumptions are

1. $(\forall x : A)(\exists! y : B(x))C(x, y)$
2. $C(x, y)$ is extensional.

The extensionality of $C(x, y)$ means that

- if x and x' are extensionally equal in A, that is, if $x =_A x'$, then x' can be substituted for x in $C(x, y)$,
- similar for every $c : B(x)(x : A)$.

(in fact, for the demonstration we only use the extensionality of the first argument of $C(x, y)$)

More generally,

- $i =_A j \rightarrow (C(i, y) \leftrightarrow C(j, y))(i, j : A)$
- $x =_{B(i)} y \rightarrow (C(i, x) \leftrightarrow C(j, y))(x, y : B(i))$

The uniqueness of y (i.e. $\exists! y$), as applied to $C(x, y)(x : A, y : B(x))$, means that [1] reads[8]

$$(\forall x : A)(\exists y : B(x))\big(C(x, y) \wedge (\forall z : B(x))(C(x, z) \rightarrow y =_{B(x)} z)\big)$$

Now that we have elucidated the assumptions, we move to the body of the proof. Let f be the function extracted from [1] by intensionality. We know that

3. $(\forall x : A)C(x, f(x))$ *true*

by the intensional Axiom of Choice. That is, we invoke the demonstration of the intensional Axiom of Choice (which uses neither uniqueness nor extensionality) and from this we take that there is a choice function, say, f.

We need to prove

4. $Ext(f)$

That is, we need to prove that

$$(\forall x, y : A)\big(x =_A y \rightarrow f(x) =_{B(x)} f(y)\big) \ true$$

Let us start by taking $a =_A b$, where $a, b : A$. We apply a and b to [3], and obtain

[8] This is one out of many ways to write an equivalent formulation in which existence and uniqueness are spelled out and accounted for.

$$C(a,f(a))\ true, \text{ and}$$
$$C(b,f(b))\ true$$

Now, according to [2], $C(x, y)$ is extensional in its first argument, and $a =_A b$ (where $a, b : A$), so we are allowed to replace the first occurrence of b in $C(b,f(b))$ with a and obtain

$$C(a,f(b))true$$

This all yields:

$$C(a,f(a))\ true, \text{ and}$$
$$C(a,f(b))\ true$$

That is we have

5. $C(a,f(a)) \wedge C(a,f(b))\ true$

Now, if we use the uniqueness condition from [1] together with $a : A$ and $f(a), f(b) : B(a)$, we obtain

6. $\big(C(a,f(a)) \wedge C(a,f(b))\big) \rightarrow f(a) =_{B(x)} f(b)\ true$

From [5] to [6], by Elimination of the implication, we obtain as required

$$f(a) =_{B(x)} f(b)\ true$$

Let us now consider the dialogical take on this version of the Axiom of Choice, as a supplement to our study in this chapter of the intensional version.

4.3.2 The Dialogical Way

From the dialogical perspective the point of uniqueness is that, without it, **P** cannot "extract" an extensional function from **O**'s concessions. In other words, when **O** challenges the extensionality of the function, **P** cannot force **O** to concede that extensionality applies to that function—even if **O** concedes that there is such a function. In order to do so **O** has to concede that there is only one element in the set $B(x)$.

The rule that provides the local meaning of the expression $Ext(f)$ is the following:

Posit	Challenge	Defence
X $!p : Ext(f)$ (where $f : A \mapsto B$)	**Y** $!L^{Ext(f)}(p) : k_i =_A k_j$ $< k_i : A, k_j : A >$	**X** $!R^{Ext(f)} f(k_i) =_B f(k_j)$

The idea underlying this rule should be clear: the play-object p is constituted by a pair: the left side and the right side of p. If **X** posited that $f : A \mapsto B$ is extensional, then the left part of p is a play-object for the extensional equality (in A)—posited by

Y—of two elements of A chosen by the challenger. The right part of p is a play-object for the extensional equality (in B)—posited by **X**—of the two functions taking as arguments the elements of A chosen by the challenger.

We will also need a rule for substitution in extensional relations (for a binary relation C):

Posit	Challenge	Defence
$\mathbf{X}\,!p : Ext(C(k_i, k_z))(k_i : A, k_z : B)$	$\mathbf{Y}\,!L_1^{Ext(C(x,y))}(p) : k_j =_A k_i$	$\mathbf{X}\,!R_1^{Ext(C(x,y))}(p) : C(k_j, k_z)$
	$\mathbf{Y}\,!L_2^{Ext(C(x,y))}(p) : k_w =_B k_z$	$\mathbf{X}\,!R_2^{Ext(C(x,y))}(p) : C(k_i, k_w)$

If **X** posited that there is a play-object p for the extensionality of the relation C, then there are two possible challenges:

- The first challenge yields a play-object for the extensional equality (in A)—posited by **Y**—of the first term of the relation with an element of A. The right part of p is then a play-object for C, posited by the defender, in which the first term of C has been substituted by the element of A that the challenger posited as being extensionally equal to the first element of the relation.
- The second challenge yields a play-object for the extensional equality (in B)—posited by **Y**—of the second term of the relation with an element of B. The right part of p is in that case a play-object for C, posited by the defender, in which the second term of C has been substituted by the element of B that the challenger posited as being extensionally equal to the second element of the relation.

We are now in position to present a dialogue for the Axiom of Choice in its extensional version with uniqueness. For the record, its fully fledged formulation is the following:

$$(\forall x : A)(\exists y : B(x))\big(C(x, y) \wedge (\forall z : B(x))(C(x, z) \to y =_{B(x)} z)\big) \to$$
$$(\exists f : (\forall x : A)B(x))\big(Ext(f) \wedge (\forall x : A)C(x, f(x))\big)$$

Compared to the intensional version studied previously in this chapter, notice the following differences:

- The antecedent is straightened so that y is unique. This means that the Opponent, who will be the one to posit the antecedent, will be committed to more and thus will have to concede more during the game.
- The consequent is augmented with $Ext(f)$. This means that, in addition to exhibiting a function f such that $C(x, f(x))$, the Proponent will have to justify that f is extensional.

Let us focus on the consequent which **P** has to endorse. In order to have a winning strategy, there are three things which **P** must do if the Opponent asks him, namely: (i) exhibit a function from A to $B(x)$ such that (ii) $C(x, f(x))$ and that (iii) f is extensional.

More precisely, the point is that the Opponent can ask for any of these and **P** must be able to defend each one successfully if he is to have a winning strategy. Since items (i) and (ii) were already at stake in the intensional version, we only present a play dealing with (iii). The two other plays which are needed to get the core of a winning **P**-strategy are almost identical to the ones we considered in Sect. 4.1.

Another difference with the intensional version is that there is an additional initial concession: in addition to $H1$-$H3$—related to the formation of expressions at stake—the Opponent concedes with $H4$ that the relation C is itself extensional. Summing up, we consider in Table 4.3 p. 83 a play for the extensional version of the Axiom of Choice with uniqueness of y, under the four initial concessions $H1$-$H4$ and in which the Opponent challenges the extensionality of f.

For the sake of concision we skipped the moves for resolution of instructions and performed the substitutions directly. This is done to some extent at the expense of clarity, but we thought that the length of the fully detailed play would also have been an obstacle to the clarity of the explanations.
Description of the play:

- Moves $H1$ to 4: The initial concessions and the thesis are posited. After that, the players choose their repetition ranks and the Opponent challenges the thesis as prescribed by the particle rule for material implication. **P** answers accordingly with move 4.
- Move 5: At this point the Opponent must choose: she can ask either for the left side or the right side of the existential posit made by **P** at move 4. Notice that the case in which she asks for the left side yields more or less the first play given for the intensional version in this chapter. Here she chooses to ask for the right side.
- Moves 6 to 8: The Proponent posits the right side of the existential, which involves the instruction $L^{\exists}(g, r)$ for the left side. The Opponent asks **P** to resolve this instruction at move 7, and **P** does so at move 8.
- Move 9: Contrary to what happened in the case of the intensional version, the right side of the consequent—posited by **P** in move 6—is now a conjunction. Thus the Opponent once more has a choice: she can ask for the left or right side of this conjunction.
  The choices available to the Opponent for her moves 5 and 7 account for the triple commitment of **P** which we explained above.
  Since we are interested in the problem of the extensionality of f, we consider the case in which **O** asks for the left side. The case in which she asks for the right side amounts more or less to the second play of the intensional version given in this chapter.
- Moves 10 and 11: The Proponent posits that f is extensional and the Opponent challenges it according to the rule we just gave.
- Moves 12 to 49: These moves describe the counterattack by which **P** will be able to make **O** posit herself the move he needs. Let us give some details:
  - Moves 12 to 15: **P** starts by making **O** choose a play-object for $B(a)$. Notice that **P** can choose $a : A$ in his move 12 because of **O**'s move 11.

Table 4.3 Play for the extensionality of f

	O			**P**	
$H1$	$C(x,y):set(x:A,y:B(x))$				
$H2$	$B(x):set(x:A)$				
$H3$	$A:set$				
$H4$	$e:Ext(C(x,y))(x:A,y:B(x))$				
				$p:(\forall x:A)(\exists y:B(x))(C(x,y)\wedge(\forall z:B(x))(C(x,z)\rightarrow y=_{B(x)}z)) \rightarrow(\exists f:(\forall x:A)B(x))(Ext(f)\wedge(\forall x:A)C(x,f(x)))$	0
1	$n:=1$			$m:=2$	2
3	$L^{\rightarrow}(p)/g:(\forall x:A)(\exists y:B(x))(C(x,y)\wedge(\forall z:B(x))(C(x,z)\rightarrow y=_{B(x)}z))$	(0)		$R^{\rightarrow}(p)/(g,r):(\exists f:(\forall x:A)B(x))(Ext(f)\wedge(\forall x:A)C(x,f(x)))$	4
5	$?_R$	(4)		$R^{\exists}(g,r)/r:Ext(L^{\exists}(g,r))\wedge(\forall x:A)C(x,L^{\exists}(g,r)(x))$	6
7	$L^{\exists}(g,r)/?$	(6)		$r:Ext(g)\wedge(\forall x:A)C(x,g(x))$	8
9	$?_L$	(8)		$L^{\wedge}(r)/s:Ext(g)$	10
11	$L^{Ext(g)}(s)/t:a=_A b < a:A,b:A >$	(10)		$R^{Ext(g)}(s)/q3:g(a)=_{B(a)}g(b)$	50
13	$R^{\forall}(g)/u:(\exists y:B(a))(C(a,y)\wedge(\forall z:B(a))(C(a,z)\rightarrow y=_{B(a)}z))$		(3)	$L^{\forall}(g)/a:A$	12
15	$L^{\exists}(u)/v:B(a)$		(13)	$?_L$	14
17	$g(a)=v:B(a)$		(3, 13, 15)	$?_{\Pi-eq}$	16
19	$R^{\exists}(u)/w:C(a,L^{\exists}(u))\wedge(\forall z:B(a))(C(a,z)\rightarrow L^{\exists}(u)=_{B(a)}z)$		(13)	$?_R$	18
21	$w:C(a,g(a))\wedge(\forall z:B(a)(C(a,z)\rightarrow g(a)=_{B(a)}z)$		(19)	$L^{\exists}(u)/g(a)$	20
23	$L^{\wedge}(w)/w_1:C(a,g(a))$		(21)	$?_L$	22
25	$R^{\forall}(g)/u':(\exists y:B(b))(C(b,y)\wedge(\forall z:B(b))(C(b,z)\rightarrow y=_{B(b)}z))$		(3)	$b:A$	24
27	$L^{\exists}(u')/v':B(b)$		(25)	$?_L$	26
29	$g(b)=v':B(b)$		(3,25,27)	$?_{\Pi-eq}$	28
31	$R^{\exists}(u')/w':C(b,L^{\exists}(u'))\wedge(\forall z:B(b))(C(b,z)\rightarrow L^{\exists}(u')=_{B(b)}z)$		(25)	$?_R$	30
33	$w':C(b,g(b))\wedge(\forall z:B(b))(C(b,z)\rightarrow g(b)=_{B(b)}z)$		(31)	$L^{\exists}(u')/g(b)$	32
35	$L^{\wedge}(w')/w'_1:C(b,g(b))$		(33)	$?_L$	34
37	$B(a)=B(b):set$		(H2)	$a=b:A$	36
39	$g(b)=v':B(a)$		(29)	$?_{Set\text{-}Eq}$	38
41	$e:Ext(C(b,g(b)))$		(H4)	$b:A,g(b):B(a)$	40
43	$R^{Ext^{C(x,y)}}(e)/q_2:C(a,g(b))$		(41)	$L^{Ext^{C(x,y)}}(e)/t:a=_A b$	42
45	$R^{\wedge}(w)/q:(\forall z:B(a))(C(a,z)\rightarrow g(a)=_{B(a)}z)$		(21)	$?_R$	44
47	$R^{\forall}(q)/q_1:C(a,g(b))\rightarrow g(a)=_{B(a)}g(b))$		(45)	$L^{\forall}(q)/g(b):B(a)$	46
49	$R^{\rightarrow}(q_1)/q_3:g(a)=_{B(a)}g(b)$		(47)	$L^{\rightarrow}(q_1)/q_2:C(a,g(b))$	48

- Moves 16-17: **P** uses the fact that **O** played moves 3, 13 and 15 to apply the Π-Equality rule. As a result, the Opponent must concede that $g(a)$ is of type $B(a)$. Notice that this action is similar to the one in the second play for the intensional version of the Axiom of Choice.
  Indeed, as we have stressed above, the extensional version is similar to the intensional one, except that it comes with stronger burdens for the two players. The point is that the reasons why $C(x, f(x))$ can be concluded are the same as in the intensional version.[9] The extensional version does not differ from the intensional one on that particular point: the extensionality of f is a supplementary burden, not a modification.
- Moves 18 to 23: The Proponent keeps on challenging until the Opponent posits $w_1 : C(a, g(a))$. Notice that because of the Opponent's moves 15 and 17, the Proponent is able to require at move 20 that the substitution $L^{\exists}(u)/g(a)$ is made in the Opponent's posit 19.
- Moves 24 to 35: The Proponent uses the same method to make **O** posit $w_1' : C(b, g(b))$. Notice that the Proponent's challenge 24 is possible because his repetition rank is 2 and, more important, because of **O**'s move 11.
- Moves 36 and 37: At this point the Proponent uses the equality $a =_A b$ conceded by **O** at move 11 to challenge $H2$. The Opponent then must posit that $B(a)$ and $B(b)$ are equal sets.
- Moves 38 and 39: Since $B(a)$ and $B(b)$ are equal sets, and since $g(b) = v' : B(b)$ has been posited at move 29, the Proponent can make **O** posit that $g(b) = v' : B(a)$.
- Moves 40 to 43: At this point the Proponent uses the extensionality of C conceded in $H4$ to make **O** posit $q_2 : C(a, g(b))$. The Proponent does so by challenging the extensionality of C at move 42: once again, this challenge by the Proponent is possible because of the Opponent's move 11.
- Moves 44 to 49: With the Opponent's posits of moves 23, 35, 39 and 43, the Proponent now has everything he needs to use the uniqueness of y. He can choose $g(b)$ for z at move 46 and then challenge the material implication by positing $q_2 : C(a, g(b))$. The Opponent must then posit the consequent and concede that $g(a)$ and $g(b)$ are extensionally equal in $B(a)$.

- Move 50: **P** can finally answer the Opponent's challenge 11 by copy-catting **O**'s move 49 in order to defend his posit that f is extensional. Since **O** has no further possible move, the Proponent wins.

[9] This is what Martin-Löf (2006) means when he writes p. 349 "*by the intensional Axiom of Choice, there exists a choice function* (...)". The idea is not that the intensional version must be assumed to demonstrate the extensional version, but the point is that we can conclude in the extensional case that there is a choice function for the same reasons that we can conclude it in the intensional case.

References

Bell, J. 2009. *The Axiom of Choice*. London: College Publications.

Bernays, P. 1930. Die Philosophie der Mathematik und die Hilbertsche Beweistheorie. *Blätter für deutsche Philosophie* 4: 326–367. (Translated in Mancosu (1998)).

Hintikka, J. 1996a. *The Principles of Mathematics Revisited*. Cambridge: Cambridge University Press.

Hintikka, J. 1996b. *Lingua Universalis versus Calculus Ratiocinator: An Ultimate Presupposition of Twentieth-Century Philosophy*. Dordrecht: Kluwer.

Jovanovic, R. Hintikka's take on the axiom of choice and the constructivist challenge. To appear in *Revista de Humanidades de Valparaiso*.

Lorenz, K. 1970. *Elemente der Sprachkritik: Eine Alternative zum Dogmatismus und Skeptizismus in der Analytischen Philosophie*. Frankfurt/Main: Suhrkamp.

Mancosu, P. 1998. *From Brouwer to Hilbert*. Oxford: Oxford University Press.

Martin-Löf, P. 1984. *Intuitionistic Type Theory*. Naples: Bibliopolis. Notes by Giovanni Sambin of a series of lectures given in padua, June 1980.

Martin-Löf, P. 2006. 100 years of Zermelo's axiom of choice: What was the problem with it? *The Computer Journal* 49(3): 345–350.

Rahman, S., and N. Clerbout. 2013. Constructive type theory and the dialogical approach to meaning. *Baltic International Yearbook of Cognition, Logic and Communication*, vol. 8. doi:10.4148/1944-3676.1077, http://dx.doi.org/10.4148/1944-3676.1077.

Sundholm, G. 1997. Implicit epistemic aspects of constructive logic. *Journal of Logic, Language and Information* 6(2): 191–212.

Sundholm, G. 2001. A plea for logical atavism. In *The Logica Yearbook 2000*, ed. O. Majer, 151–162. Prague: Filosofia.

Tait, W. 1994. The law of excluded middle and the axiom of choice. In *Mathematics and Mind*, ed. A. George, 45–70. New York: Oxford University Press.

Chapter 5
Building a Winning P-strategy Out of a CTT Demonstration

In this brief chapter we will consider the other direction of the equivalence result between the valid fragments of the CTT framework and the dialogical framework. That is, we will show that if there is a CTT demonstration for φ then there is a winning **P**-strategy in the dialogical game for φ.

The proof, quite unsurprisingly, rests on developing a translation procedure which is the converse of the one in Chap. 3. That is, we will present a procedure transforming a given CTT demonstration and we will show that the result is the core of a winning **P**-strategy—which is then expanded to a fully fledged winning strategy.

A core is expanded to a full strategy by adding branches accounting for variations in the order of the moves of the other player and in the play-objects he chooses. We will not give the specifics of that particular operation because it does not hold any difficulty and they have been given with details elsewhere (Clerbout 2014a, b). We would rather focus on the way the initial CTT demonstration is transformed and on the proof that the result is the core of a winning **P**-strategy.

For the latter, we need to prove that the transformation results in a tree in which each branch represents a play won by **P**. In other words in which each branch represents a legal sequence of moves ending with a **P**-move or with **O** positing $\perp$. We also need to check that the tree has all the necessary information to be a core which can be expanded to a full strategy. That is to say, we must make sure that no possible play for **O** is ignored, excepting those varying in the order of the moves or the names of the play-objects.

N. Clerbout and S. Rahman, *Linking Game-Theoretical Approaches with Constructive Type Theory*, SpringerBriefs in Philosophy,
DOI 10.1007/978-3-319-19063-1_5

5.1 Transforming CTT Demonstrations

Before we get there we need to design a transformation procedure. We will start with an informal description of the task and of the ideas underlying the procedure. Then we will give the detailed algorithm.

5.1.1 Guidelines

In general there are two main difficulties such a procedure must overcome:

1. CTT is not an interactive-based framework. In particular the notions of players, challenges and defences do not make any sense in CTT.
2. The progression of a CTT demonstration differs quite greatly from the progression of a dialogical core. Most notably, the production of ramifications on the one hand and the order of expressions on the other hand do not match in the two approaches.

These are just descriptions of the fundamental differences between a CTT demonstration and a dialogical core. There are obviously many other aspects which our translation method must take into account. Let us give further explanations on the topics on which the desired transformation procedure must operate.

5.1.1.1 From CTT Judgements to Dialogical Posits

To begin with we need to enrich the CTT demonstration with the players' identities. We need for that a way to figure out which expressions are posited by which player. In fact, there is a subtlety in this process because some steps in a CTT demonstration may be associated with both players: see "Identical posits by the two players" below for more on this. But the general idea underlying the process is otherwise quite simple. The starting point is that the conclusion of the CTT demonstration is to be posited by the Proponent because it is the expression at stake. In a dialogue, that is the thesis. Moreover, the hypotheses of the demonstration, that is, the undischarged assumptions that may occur at the leaves of the CTT demonstration, correspond to initial concessions made by the Opponent.

From there, it is quite easy to associate the other steps in the CTT demonstration with players by using the correspondences between the CTT and dialogical rules which we used in Chap. 3. By means of illustration, suppose some step in the CTT demonstration has been associated to player **X** and suppose that the expression results from an application of the $\to$ Introduction rule. Then the assumption discharged by applying the Introduction rule is to be associated to player **Y** (it will occur in the core as the challenge by **Y**) and the expression immediately preceding the inference line is to be associated to player **X** (it will occur in the core as the defence).

5.1.1.2 Identical Posits by the Two Players

Because the CTT framework is not based on interaction, it does not distinguish between the two players. A consequence is that having an expression occurring once as an **O**-posit and once as a **P**-posit does not make any sense in the CTT perspective. In the dialogical framework however, this is not only possible but sometimes crucial because, as we have seen, a form of copy-cat often underlies winning strategies for the Proponent. The clearest and most basic example is the case of the Modified Formal Rule $SR3$ on elementary expressions.

The point is that a CTT demonstration may very well feature expressions occurring only once, while two instances (or more) would be needed for a dialogical demonstration, that is, for the construction of a dialogical core. Elementary expressions associated to the Proponent, and which do not result from the application of the $\bot$ Elimination rule, are one example. More generally, an expression may be used in a demonstration when applying the two kinds of rules: for example it can be used first when applying an Elimination rule and later on when applying an Introduction rule. In such cases, this expression is likely to occur as posited by the two players in a dialogical core (intuitively, this is because of the correspondence between Elimination rules and **O**-applications of rules on the one hand and Introduction rules and **P**-applications of rules on the other hand). This results in adding occurrences of expressions, but as posited by a different player.

5.1.1.3 Dialogical Instructions and Play-Objects

Next we need to account for the difference between CTT proof-objects on the one hand, and dialogical play-objects and instructions on the other hand. More precisely, we need to go from the CTT perspective on applications of rules to the dialogical perspective. In the CTT framework, applications of rules manifest themselves by specific operations defining the way proof-objects are obtained from other proof-objects. In the dialogical approach, meaning explanations are given in terms of play-objects and instructions at the other (preliminary) level of plays in which interaction prevails over the set-theoretic operations.

To perform this change of perspective, we start by substituting an arbitrary play-object p for the proof-object in the conclusion of the demonstration: in other words, we choose an arbitrary play-object for the thesis of the dialogical core we are building. Also, if relevant, we substitute play-objects for proof-objects in the expressions corresponding to initial concessions by the Opponent.

From there, it is a trivial matter to replace the other proof-objects occurring in the demonstration with the appropriate dialogical instructions. We simply look which rule is applied to know which subscript must be associated to the letters L and R which will result in a proper dialogical instruction. For example, an instruction of the form $L^{\wedge}(...)$ (or $R^{\wedge}(...)$) is substitued for the proof-object of the conclusion resulting from an application of the $\wedge$ Elimination rule in the initial CTT demonstration.

We also need to account for the dialogical process of Resolution of Instructions (structural rule $SR4.1$). We do so as we replace CTT proof-objects with dialogical instructions: every time we determine the dialogical instruction replacing the CTT proof-object, we also choose a play-object resolving the instruction. As a result, an expression "$\alpha : \chi$", where α is a proof-object, will be replaced by an instruction of the form "$I : \chi$" where I is an instruction. Immediately after that another version of the same posit is added in the structure, but with a play-object instead of the instruction I. The reason for this is that we can progressively replace proof-objects with simple instructions relative to play-objects, instead of having embedded instructions getting more and more complicated.[1]

5.1.1.4 Adding Questions

At this point we have obtained a tree-like structure featuring a lot of expressions which differ only by the player identity and/or maybe by the instructions and play-object. From the CTT perspective, these look like irrelevant and unnecessary repetitions: such a structure does not make sense without the notion of interaction between two players.

Still, some aspects are missing to read the structure at hand in terms of interaction. To put it simply, the structure lacks challenges consisting in questions. For example, that two expressions $\mathbf{X}\,!I : \chi$ (for some instruction I) and $\mathbf{X}\,!p : \chi$ (for some play-object p) follow each other in the structure does not make sense until the question $\mathbf{Y}\ I/?$ is placed between them: only then can we speak of an interaction in which **Y** asks **X** for the resolution of the instruction I and **X** chooses p for the resolution. Similarly with other questions such as $?_\vee$, $?_L$, $?_R$, etc., depending on the rule at stake.

The next step in the translation procedure is therefore to include questions in the relevant way so that one can accurately speak of interaction through the application of dialogical rules. However, the result still cannot be called a dialogical core. For that we need to overcome the difference in the production of ramifications between the CTT framework and dialogical strategies.

5.1.1.5 Rearranging the Branches and Order of the Moves

For the clarity of the explanation, recall that we are dealing with a tree-like structure written "upside-down", that is, where the root of the tree (the conclusion of the demonstration we started with) is at the bottom and the leaves are at the top.

The most important transformation that remains is reorganising the tree at hand so that we obtain a good candidate for a core of a winning **P**-strategy. This means we aim to obtain a tree in which branches are linear representations of plays in such a way

[1] Suppose for example that we have introduced an instruction $L^\vee(p)$. If we do not decide immediately for a play-object, say q, to resolve this instruction, then the next instruction will be of the form $I(L^\vee(p))$ instead of the simpler $I(q)$—for some I.

that ramifications represent choices of the Opponent between different moves (since we are interested in **P**-strategies). The CTT framework distinguishes between rules applied to one or more expressions. In the latter case, a ramification is produced but not in the former case. But since there is no notion of interaction and strategy (in the game-theoretical sense) in Constructive Type Theory, it is obvious that ramifications may not correspond to differences due to possible choices by a player, taken into account in a strategy for his adversary.

A typical example are the differences between the CTT Elimination rules for material implication and universal quantification on the one hand, and their dialogical counterpart on the other hand. In CTT these rules have (at least) two premisses: first the complex expression and second a judgement of the form $a : A$ when A is the antecedent or the set which is quantified over. Each of these two premisses opens a branch in the demonstration. But in a dialogical game, one is the posit and the other is the challenge against it. Consequently they occur in the same play and hence, in the same branch of a strategy. Notice that the same difference occurs with other rules, most notably with equality rules.

The goal in this step of the transformation is thus to reorganise the tree in order to overcome these differences. We must also take some additional precautions (such as adding the choices of repetition ranks) so that the branches in the new tree do indeed represent plays.

We shall stop the general explanations here. All the details are given in the full description of the translation algorithm given below. Let us simply mention here that the procedure is meant to obtain the core of a winning **P**-strategy after all these modifications. This is something that must be proved, which will be done in Sect. 5.2.

5.1.2 The Procedure

We will now give the specifics of the translation procedure briefly described above. Let us suppose a CTT demonstration D of an expression E under a set H of hypotheses.

A—From judgements to posits First we enrich the initial demonstration with player identities and the posit sign !

A1. Rewrite the conclusion E as $\mathbf{P}\,!E$. Then, for every $h \in H$ occurring as a leaf of D, rewrite h as $\mathbf{O}\,!h$. Go to A2.

A2. Scan D bottom-up. When there is no unused expression left, go to A3. Otherwise, let E_1 be the (leftmost[2]) unused expression $\mathbf{X}\,!E_1$. Then,

1. If $\mathbf{X}$ is $\mathbf{O}$ and E_1 results in D from applying an Introduction rule, then insert $\mathbf{P}\,!E_1$ as the conclusion of the rule preceding $\mathbf{O}\,!E_1$. Consider the latter as used and go back to A2.

[2]This accounts for the fact that D may have several branches.

2. If **X** is **P** and E_1 results in D from applying an Elimination rule other than for $\vee$ or Σ, then insert $\mathbf{O}\,!E_1$ as the conclusion of the rule preceding $\mathbf{P}\,!E_1$. Consider the latter as used and go back to A2.
3. Otherwise use the correspondences between CTT and dialogical rules given in Chap. 3 to rewrite the expressions allowing the application of the rule with the adequate player.[3]

   In doing so, observe the following constraints:

   - an expression can be labelled as a **P**- *and* an **O**-posit,
   - each player can be assigned at most once to an expression.

Consider the expression as used and go back to A2.

A3. Scan the demonstration at hand. For each elementary posit by the Proponent which has no counterpart by the Opponent, apply one of the following,

(1) If it is the result of an application of the $\bot$ Elimination rule, then remove it.
(2) If there is no corresponding **O**-posit, then insert one immediately below the Proponent's posit, and insert the expression $\mathbf{P}\, sic(n)$ at the leaf of the current branch.

Then go to A4.

A4. If there are leaves with the double label $\mathbf{O}\,!/\mathbf{P}\,!$, separate them into two expressions such that the Proponent's posit is placed as the leaf. Go to B.

B—Instructions and play-objects. Next we substitute play-objects and dialogical instructions for CTT proof-objects. This is done in the following way.

B1. In the conclusion $\mathbf{P}\,!E$, replace the proof-object with an arbitrary play-object p. Then, for each initial concession $\mathbf{O}\,!h$ occurring at a leaf of the demonstration, substitute, if relevant, an arbitrary play-object for the proof-object. Consider these expressions as treated and go to B2.

B2. Scan the demonstration bottom-up. If there is no expression left untreated, go to C. Otherwise take the leftmost expression $\mathbf{X}\,!E_2$ with a play-object which has not been treated so far, and

- Use the correspondences between CTT and dialogical rules given in Chap. 3 to substitute the adequate instructions for the proof-object(s) in the premiss (premisses) of the rule whose application results in $\mathbf{X}\,!E_2$.

[3] Let us give a simple illustration instead of recalling the correspondences. Assume the situation is

$$\frac{\alpha : (\Pi x : A)\varphi \quad a : A}{\mathbf{O}\,!...}\,\Pi E$$

Then the result is

$$\frac{\mathbf{O}\,!\alpha : (\Pi x : A)\varphi \quad \mathbf{P}!a : A}{\mathbf{O}\,!...}\,\Pi E.$$

- For each instruction introduced that way, copy the expression at stake, replacing the instruction by an arbitrary play-object. Place the version with the play-object immediately above the expression with the instruction.
- Consider **X** $!E_2$ as treated and go back to B2.

C—Adding questions. Scan the demonstration and identify the applications of rules for which the dialogical counterpart features a question. For each expression understood as a defence according to such a rule, add the corresponding challenge performed by the adversary immediately below the expression.

Go to D.

D—Move the Opponent's initial concessions. Consider each leaf of the demonstration at hand which is an initial concession by the Opponent—that is, an undischarged assumption of the initial demonstration D which has been identified as an Opponent's move. Remove it and place it below the conclusion **P** $!E$. In case of multiple occurrences, keep only one occurrence.

Go to E.

E—Removing non-dialogical splits. Scan the demonstration top-down. Going from the left to the right, check each point where two different branches join. Depending on what the case may be, apply one of the following,

(1) If the ramification is such that the two branches are opened by two **O**-posits relevant for a rule dealing with a logical constant, then ignore and proceed downwards.
(2) Otherwise, "cut" one and "paste" it above the other one, according to the following convention:
   - If both branches have a **P**-move as the leaf, or if both have an **O**-move as the leaf, then pick any one of the branch to be cut and pasted,
   - Otherwise pick the one with a **P**-move at the leaf to be cut and pasted.

Go to F.

F—Reordering the nodes. Scan the tree structure at hand bottom-up. Starting from the thesis **P** $!E$, change the order of the expressions according to the following conditions,

- Each **O**-move is a reaction—as specified by the dialogical rules—to the **P**-move placed immediately below,
- A question or a posit which is a challenge always occurs before (i.e. closer to the root) a defence reacting to it.
- Ramifications are preserved so that each branch is opened with an **O**-move as a reaction to a **P**-move which is immediately below.

Go to G.

G—Adding ranks. Insert an expression **O** $\mathtt{n} := \mathtt{1}$ immediately above the thesis **P** $!E$. Then insert an expression **P** $\mathtt{m} := \mathtt{k}$ above the one just inserted. Choose $\mathtt{k}$ to be the biggest number of times a given rule is applied by **P** to the same expression in the tree.

The procedure stops.

5.2 Adequacy of the Algorithm

We have given the algorithm transforming a CTT demonstration. It should be obvious from the general ideas at the beginning of the previous section that the algorithm is designed to make the differences disappear between CTT demonstrations and dialogical cores of winning strategies. It remains to show that the algorithm indeed does so, in other words that applying the algorithm to a given a CTT demonstration results in the core of a winning **P**-strategy.

To be more specific, the point is to show that the result of applying the algorithm to a CTT demonstration is a tree in which,

(i) Each branch represents a play: the sequence of moves in each branch complies with the game rules,
(ii) Each play in the tree is won by the Proponent,
(iii) The tree describes all the relevant alternatives for a core. In other words: there is no significantly different[4] course of action for **O** that would be disregarded in the resulting tree.

Proposition 5.1 *Each branch in the resulting tree represents a play.*

We need to show that in each branch the sequence of moves complies with the rules of dialogical games of Chap. 2.

Proof Because the translation observes a correspondence between CTT rules and dialogical particle rules,[5] we simply need to check that the dialogical structural rules are observed. We leave the Winning Rule aside for now since it is the topic we address next in Proposition 5.2.

So for the Starting Rule $SR0$, steps D and G of the algorithm ensure that every sequence of moves in the tree starts with the initial concessions of the Opponent, which are followed by the thesis posited by the Proponent and then by the choices of repetition ranks.

As for the Intuitionistic Development Rule $SR1i$, step F of the algorithm guarantees that each move following the repetition ranks in a sequence is played in reaction to a previous move. The condition in step F according to which **O**-moves immediately follow the **P**-move to which it is a reaction ensures that the intuitionistic restriction of the *Last Duty First* is observed.[6] Moreover the choice of the repetition

[4]By significantly different we mean other than relative to the order of the Opponent's moves, or the choice of play-objects to replace instructions.

[5]And given that step C of the algorithm has been used to insert questions.

[6]This is known since Felscher (1985). See also more details in Clerbout (2014b).

ranks prescribed by step G ensures that the players do not perform unauthorised repetitions.

As for the Modified Formal Rule $SR3$, no challenge against an elementary posit made by **O** is added when applying the algorithm. Moreover, in the case of elementary posits made by **P**, step A3 of the procedure ensures that, if needed, a corresponding posit by **O** and the adequate challenges and defences are added.

As for the rules related to the Resolution of Instructions ($SR4.1$ and $SR4.2$), step B of the algorithm (in combination with step C introducing questions) guarantees that instructions are resolved according to the structural rules.

As for the rule $SR2$ related to formation dialogues, recall it should be ignored because we are focusing on the valid fragment of CTT in which verifying the well formation is assumed successful.

Now that we have established that the branches represent plays because they comply with the dialogical rules, we must assess the situation in relation to victory and show that:

Proposition 5.2 *Each branch of the resulting tree represents a play won by* **P**.

Proof We must check that the leaf of each branch is either:

1. an elementary posit by **P**,
2. the posit by **O** of $\bot$,
3. a **P**-move "$sic(n)$" for some move numbered n.

But all this is guaranteed by steps A3, E and F of the algorithm.

Finally, it remains to show that the tree describes all the relevant courses of actions for the Opponent underlying the core of a **P**-strategy:

Proposition 5.3 *There is no* **P***-move in the tree remaining unanswered by* **O** *while some rule would allow it.*

Proof We know from the initial demonstration D and steps A1–A4 of the algorithm that every posit made by the Proponent in the resulting tree occurs as the result of an Introduction rule, of the Elimination rule for the Σ operator or of the Elimination rule for disjunction. In the case of complex posits, the correspondence with dialogical particle rules together with the addition of questions via step C of the algorithm ensure that they are challenged and that when they are themselves played as challenges they are answered. In the case of elementary posits, we know from the proof of Proposition 5.1 that they are challenged if the Opponent can.

Moreover, all the possible challenges allowed by the particle rules are covered by the CTT rules they correspond to. For this reason, the only remaining possible variations left to the Opponent are the order of her moves and the choice of play-objects for the Resolution of Instructions (as specified by the structural rules $SR4.1$ and $SR4.2$). But these variations are the ones which are not relevant to build the *core* of a **P**-strategy. In other words the correspondence between CTT rules and the particle rules ensure that the starting demonstration D already contains the variations which are relevant for a core of **P**-strategy.

The adequacy of our translation procedure, which amounts to the second direction of the equivalence result we stated at the beginning of this study, is then a direct consequence of Propositions 5.1–5.3:

Corollary 5.1 *The result of applying the algorithm of Sect.* 5.1.2 *to a CTT demonstration is the core of a winning* **P***-strategy.*

References

Clerbout, N. 2014a. First-order dialogical games and tableaux. *Journal of Philosophical Logic* 43(4): 785–801. [Published first online 2013. doi:10.1007/s10992-013-9289-z].

Clerbout, N. 2014b. *La Sémantique Dialogique. Notions Fondamentales et Éléments de Metathéorie*. London: College Publications.

Felscher, W. 1985. Dialogues as a foundation for intuitionistic logic. In *Handbook of Philosophical Logic*, vol. 3, ed. D. Gabbay and F. Guenthner, 341–372. Dordrecht: Kluwer.

Chapter 6
Conclusions and Work in Progress

Developing a dialogical approach to CTT is still only beginning and many open issues have yet to be tackled in order to assess its fruitfulness. However, before these issues could be tackled, the dialogical approach to CTT had to be shown faithful to the standard CTT formulation. That task has just been accomplished, so let us now briefly mention some paths of research in which progress is currently being made:

1. **Harmony, Π and Σ-Equality Rules, and Copy-Cat**. In Sect. 2.3, we briefly mentioned how the dialogical approach to CTT opens new paths in studies on the notion of harmony. As we have argued, a reason for this is that player-independence allows a version of Π and Σ-Equality which does not dichotomise Introduction and Elimination as in the case of CTT.
   However, we have also noticed that these Equality rules seem to share a close relationship with copy-cat strategies. Obviously these are not available in a framework such as CTT in which meaning is not linked with interaction. Further studies will investigate the possibility to replace the dialogical version of Equality rules with an analysis using the more game-theoretical notion of copy-cat. One of the challenges is that explicit reasons for why copy-cat moves are possible are not usually given. In the case we are concerned with, it amounts to failing to account for the equalities justifying copy-cat moves: the challenge is thus to come up with an analysis making the best of the implicit (by copy-cat) and explicit (in terms of equalities) approaches.
2. **The Meaning of Conversations**. In his book *The Interactive Stance*, Ginzburg (2012) stresses the utmost importance of taking conversational (interactive) aspects into account in order to develop a theory of meaning, in which meaning is constituted during the interaction. In order to implement such a theory of meaning Ginzburg uses CTT in which the so called "metalogical" rules constituting meaning are explicitly imported into the object language. Moreover,

N. Clerbout and S. Rahman, *Linking Game-Theoretical Approaches with Constructive Type Theory*, SpringerBriefs in Philosophy,
DOI 10.1007/978-3-319-19063-1_6

Ginzburg designed some kind of language games called dialogical-gameboards that are meant to capture the dynamic aspects of everyday dialogues.
Now, if we take seriously the claim that meaning is constituted by and within interaction, then we expect the semantics of the underlying logical elements to be also understood dialogically. In this context, a dialogical approach to Constructive Type Theory provides both a dialogical framework for the underlying logic and a natural link to the dialogical gameboards. Rahman (2014) has started tackling this issue but a full development is still to be worked out.

3. **R. Brandom's Pragmatic Rationalism and the Dialogical Perspective**. Brandom's pragmatic rationalism links meaning with the task of asking and giving reasons in the context of commitments triggered by assertions. The philosophical approach of Brandom seems to be close to the formal developments of Ginzburg mentioned above. However the dialogical perspective places meaning at a lower level: the level of all possible plays displayed in the extensive form of the underlying game. The claim of the dialogical perspective is that understanding involves knowing how to play and not only how to win. That is to say, the dialogical approach provides a level of analysis where use (meaning) is not considered under the strong point of view of (correct) inference. Rather, the stronger notion of use as performing inferences is reached at the strategical level based on the lower level of use within plays. Thus, it seems important to examine Brandom's approach in the context of a theory in which meaning is constituted by the tree of all possible plays triggered by a main posit.
4. **Modal Epistemic Logic and Belief Revision.** In the context of CTT, the variable in a hypothetical such as $p(x) : P(x : S)$ represents an unknown element of S that can be instantiated by some s when the required knowledge is available.[1] Thus, in this framework, instantiating the unknown element x by some s known to be a fixed (but arbitrary) element of S describes the passage from belief to knowledge. Using the current terminology of epistemic logic as an analogy—in the style of Hintikka (1962)—we say that a judgement of the form $x : S$ expresses belief rather than knowledge. For this transition to count as a transition to knowledge, it is not only necessary that $s : S$, but it is also necessary that the proof-object s is of the adequate sort.[2] In other words, we also need to have the definition $x = s : S$. This definition of x can be called an *anchoring* of the hypothesis (belief) S in the actual world.[3] The result of this anchoring process thus yields $p(x = s) : P(s : S)$. In fact, after some seminal work of Ranta (1991) there are ongoing developments by Primiero (2008, 2012) on applying CTT to belief revision. However, dynamic aspects provided by game-theoretical approaches—in which knowledge acquisition is depicted as resulting from interaction—has not been considered; and the modal formalisations of belief revision have not been studied in this framework yet either. The pending task is thus to work out

[1]Cf. Granström (2011, pp. 110–112). In fact, Chap. V of Granström (2011) contains a thorough discussion of the issue.

[2]Cf. Ranta (1994, pp. 151–154).

[3]Ranta (1994, p. 152).

the ways to combine the CTT formulation of modal logic with the dialogical approach.

5. **Constructive Foundations of Logics and Mathematics**. Prawitz's (1965, p. 33) points out that his normalisation procedure is based on the inversion principle of Lorenzen's operative logic (1955). Dialogical logic arose as a reply to some problems of operative logic (see Lorenz 2001). Normalisation has already been applied to constructive logic and mathematics so the question now is to know what the dialogical interpretation of normalisation is and what this interpretation can add.
6. **Intensional and Extensional Identity**. As already mentioned in the introduction, in some recent work Dybjer (2012) proposed to study the relation between the intensional and the extensional notions of identity defined in CTT by game-theoretical means. Once more, in this context, an approach to CTT that is already game-theoretical seems to be desirable.

References

Dybjer, P. 2012. Tests, games, and Martin-Löf's meaning explanations for intuitionistic type theory. Slides of a talk given at the congress logic and interactions, Marseille (France).

Ginzburg, J. 2012. *The Interactive Stance: Meaning for Conversation*. Oxford: Oxford University Press.

Granström, J. 2011. *Treatise on Intuitionistic Type Theory*. Dordrecht: Springer.

Hintikka, J. 1962. *Knowledge and Belief: An Introduction to the Logic of the Two Notions*. Cornell: Cornell University Press.

Lorenz, K. 2001. Basic objectives of dialogue logic in historical perspective. In *New Perspectives in Dialogical Logic*, ed. S. Rahman and H. Rückert, 255–263. New York: Springer (*Synthese* 127(1–2): special volume).

Lorenzen, P. 1955. *Einführung in die Operative Logik und Mathematik*. Berlin: Springer.

Prawitz, D. 1965. *Natural Deduction: A Proof-Theoretical Study*. PhD thesis, Stockholm: Almqvist & Wiksell.

Primiero, G. 2008. *Information and Knowledge*. Dordrecht: Springer.

Primiero, G. 2012. A contextual type theory with judgemental modalities for reasoning from open assumptions. *Logique et Analyse* 220: 579–600.

Rahman, S. 2014. On hypothetical judgements and Leibniz's notion of conditional right. In *Legal Reasoning and Logic. Past & Present Interactions*, ed. A. Armgardt, P. Canivez, and S. Chassagnard-Pinet. Dordrecht: Springer.

Ranta, A. 1991. Constructing possible worlds. *Theoria* 57(1–2): 77–99.

Ranta, A. 1994. *Type-Theoretical Grammar*. Oxford: Clarendon Press.

Errata to: Linking Game-Theoretical Approaches with Constructive Type Theory

Nicolas Clerbout and Shahid Rahman

Errata to: N. Clerbout and S. Rahman, *Linking Game-Theoretical Approaches with Constructive Type Theory*, SpringerBriefs in Philosophy, DOI 10.1007/978-3-319-19063-1

Due to miscommunication during the production process, the following items were not corrected prior to the publication of the book.

Page 6, l.5: "family of sets".

Page 7, rule +eq2 (before the paragraph starting with "where i and j are two new..."): $D(i(a)), (x)d(x), (y)e(y))$ should be $D(\boldsymbol{j}(\boldsymbol{b})), (x)d(x), (y)e(y))$.

Page 14, table representing a play: the text in the "P" column should be centered.

Page 16, last paragraph "The next rule is not a formation rule per se..." → **should be after table 2.2 on page 17**.

The online version of the original book can be found under
DOI 10.1007/978-3-319-19063-1

N. Clerbout (✉)
Instituto de Filosofía; CDHACS, Universidad de Valparaíso, Valparaíso, Chile
e-mail: nicolas.clerbout@gmail.com

S. Rahman
UMR-CNRS 8163: STL, University of Lille III, Villeneuve d'Ascq, France
e-mail: shahid.rahman@univ-lille3.fr

N. Clerbout and S. Rahman, *Linking Game-Theoretical Approaches with Constructive Type Theory*, SpringerBriefs in Philosophy,
DOI 10.1007/978-3-319-19063-1_7

Page 32, first table (top of the page): **the vertical lines of the table should be continuous, without breaks**.

Section 2.5, pp. 34–36: *Oxy* should be ***O*(*x*,*y*)**.

Page 35, Table 2.6: in the line corresponding to move IV, *Bxy* : *set* should be ***B*(*x*,*y*): *set*(*x* : *M*, *y* : *D*)**.

Page 47, paragraph called "Explanations on the final stage", l.6: "by dialogical rules are no**t** the same as".

MIX
Papier aus verantwortungsvollen Quellen
Paper from responsible sources
FSC® C105338

If you have any concerns about our products,
you can contact us on
ProductSafety@springernature.com

In case Publisher is established outside the EU,
the EU authorized representative is:
Springer Nature Customer Service Center GmbH
Europaplatz 3, 69115 Heidelberg, Germany

Printed by Libri Plureos GmbH
in Hamburg, Germany